AF845563

De Gruyter Expositions in Mathematics 58

Editors
Victor P. Maslov, Moscow, Russia
Walter D. Neumann, New York, USA
Markus J. Pflaum, Boulder, USA
Dierk Schleicher, Bremen, Germany
Raymond O. Wells, Bremen, Germany

Oleg K. Sheinman

Current Algebras on Riemann Surfaces

New Results and Applications

De Gruyter

Mathematics Subject Classification 2010: 17B65, 17B66, 17B67, 17B70, 17B80, 17B81, 14H70, 37J35, 81T40.

ISBN 978-3-11-026396-1
e-ISBN 978-3-11-026452-4
ISSN 0938-6572

Library of Congress Cataloging-in-Publication Data

A CIP catalog record for this book has been applied for at the Library of Congress.

Bibliographic information published by the Deutsche Nationalbibliothek

The Deutsche Nationalbibliothek lists this publication in the Deutsche Nationalbibliografie; detailed bibliographic data are available in the internet at http://dnb.dnb.de.

© 2012 Walter de Gruyter GmbH & Co. KG, Berlin/Boston

Typesetting: PTP-Berlin Protago-TEX-Production GmbH, www.ptp-berlin.eu
Printing and binding: Hubert & Co. GmbH & Co. KG, Göttingen
♾ Printed on acid-free paper

Printed in Germany

MIX
Papier aus verantwortungsvollen Quellen
FSC www.fsc.org FSC® C016439

www.degruyter.com

To my beloved parents Karl and Cécile Sheinman,
my wife Tanya,
and our children Lena and Lenya

Preface

The present book is an introduction into the new and fast developing field on the crossroads of infinite-dimensional Lie algebra theory and contemporary mathematical physics. It presents the theory of Krichever–Novikov algebras, Lax operator algebras, their interaction, elements of their representation theory, relations to moduli spaces of Riemann surfaces and holomorphic vector bundles on them, to Lax integrable systems and Conformal Field Theory. The previous book ([68], in Russian) by the author of this volume is focused on Krichever–Novikov algebras and presents them in more detail. The present book also treats Lax operator algebras and their fruitful applications to integrable systems and Conformal Field Theory. It presents the most essential results of [68] as well.

In the 1960s to the 1980s the theory of infinite-dimensional Lie algebras went through the period of rapid growth related to the appearance of affine Kac–Moody and Virasoro algebras, which combined obvious continuity with regard to the classical theory, and success in applications.

During the same period there was a formation of two important disciplines of mathematical physics: the theory of integrable systems and the 2-dimensional Conformal Field Theory (2D CFT). Both are characterized by deep penetration of the technique of Riemann surfaces into the theoretical physics.

The discovery[1] of the role of Kac–Moody and Virasoro algebras as algebras of formal gauge and conformal symmetries of CFT, and an important tool of integration of nonlinear equations, had connected together the Lie algebra theory with those two disciplines. Nevertheless, it was clear from the very beginning that the main objects of CFT are of a complex analytical nature, and the symmetries should also be analytic. Transition from formal to geometrical symmetries has led to the current and vector field algebras on Riemann surfaces.

In 1987, the investigations in the soliton theory and CFT led I. M. Krichever and S.P. Novikov to consider central extensions of current and vector field algebras on Riemann surfaces (regarded to as complex curves with marked points) which are now called Krichever–Novikov algebras. In 2001 I. M. Krichever proposed his theory of Lax operators with the spectral parameter on a Riemann surface [33]. This theory goes back to I. M. Krichever and S. P. Novikov's technique of finding high rank finite-zone solutions to Kadomtsev–Petviashvili and Shrödinger equations [24,25]. In 2006, I. M. Krichever and the author found out that these Lax operators form an associative

[1] This was the work of many remarkable mathematicians and physicists. Only some of them are mentioned in this book in connection with concrete questions.

algebra [35], and constructed their orthogonal and symplectic analogs which form Lie algebras. They were called Lax operator algebras.

Affine Krichever–Novikov and Lax operator algebras have the same associated graded object, namely (non-twisted) loop algebras, but they are of a quite different mathematical and physical nature. They possess an important property of almost grading which is weaker than grading but stronger than filtration. It leads to many consequences; in particular, it enables one to construct analogs of highest weight representations. Affine (non-twisted) Kac–Moody algebras belong to both new classes of Lie algebras, and as such correspond to a Riemann sphere with two marked points, zero and infinity.

Krichever–Novikov algebras have numerous relations to fundamental problems of geometry, analysis, and mathematical physics. Introduced as algebras of the Fourier–Laurent series on Riemann surfaces they appear as a part of harmonic analysis. Their representation theory is closely related to holomorphic vector bundles on Riemann surfaces and Lax integrable systems (see Chapters 2 and 6 of this volume).

There is a fundamental relation, based on the Kodaira–Spencer theory, between vector field algebras on Riemann surfaces and moduli spaces of Riemann surfaces with marked points (Chapter 3). Briefly, it is as follows. The Kuranishi tangent space to the moduli space of Riemann surfaces with marked points and fixed jets of local coordinates is isomorphic to the direct sum of certain homogeneous subspaces of the Krichever–Novikov vector field algebra. It is in the same row with the classical description of the tangent space in terms of tensors on Riemann surfaces (Beltrami differentials, quadratic differentials) going back to Teichmüller, Alfors, Bers. In the situation with one marked point, the Lie algebraic description originates in the work due to Koncevich (by means of the Virasoro algebra). The two-point situation was already investigated by Krichever and Novikov, Grinevich, and Orlov by means of Krichever–Novikov algebras.

The relation between Krichever–Novikov algebras and moduli spaces underlies their applications to 2D CFT. The understanding of 2D CFT as a projective flat connection on a moduli space of Riemann surfaces goes back to A. Polyakov [38], and is clearly formulated by Friedan and Shenker [8]. The Wess–Zumino–Novikov–Witten (WZNW) theory is a 2D conformal field theory in which the momentum-energy tensor is expressed via currents by Sugawara construction. In this kind of theory, an essential role is played by a-periods of currents which physicists call zero modes. As D. Bernard, one of the founders of CFT, stressed: "The zero modes contain almost all nontrivial information concerning a WZW model". Zero modes are elements of Krichever–Novikov current algebras and do not belong to loop algebras. This is one of main reasons for using Krichever–Novikov algebras in CFT. In Chapter 3 we construct the high-genus Knizhnik–Zamolodchikov equations by means of Krichever–Novikov algebras.

An important tool of CFT, namely the Sugawara representation for Krichever–Novikov algebras, is constructed in Chapter 2. At the "physical level of rigor" the last result was obtained in [4]. The mathematically rigorous result was obtained in [51].

Lax operator algebras are attributed to a quite different field of mathematical physics, namely to nonlinear integrable systems. The applications of current algebras to the theory of Lax equations have a long history. They were initiated in the works of I. Gelfand, L. Dikii, I. Dorfman, A. Reyman, M. Semenov-Tian-Shanskii, V. Drinfeld, V. Sokolov, V. Kac, and P. van Moerbeke. Basically these applications are related to Kac–Moody algebras which appear quite naturally in the context of Lax equations with *rational* spectral parameter. In [33] the theory of conventional Lax and zero curvature representations with rational spectral parameter was generalized to the case of algebraic curves Σ of arbitrary genus g. Such representations arise in several ways in the theory of integrable systems; cf. [25] where a zero curvature representation of the Krichever–Novikov equation is introduced, or [33] where a field analog of the Calogero–Moser system on an elliptic curve is presented. Lax operator algebras appear as the corresponding generalization of Kac–Moody algebras. They are defined in terms of Tyurin parameters originating in the theory of holomorphic vector bundles. There is a remarkable phenomenon: the constraints on Tyurin parameters following from the requirement that the space of Lax operators was closed with respect to their Lie bracket simultaneously provide sufficient conditions of Hamiltonianity of the corresponding Lax equations. We treat this phenomenon in Chapter 5 of the present book.

A similar technique enables us to assign a kind of CFT to every Lax integrable system of the type considered here. Thus we obtain a relation between Lax integrable systems and Knizhnik–Zamolodchikov connection. Mathematically this is a projective unitary representation of the Poisson algebra of classical observables of the Lax system by Knizhnik–Zamolodchikov equations. Physically, this is a prequantization of the Lax integrable system. We treat these questions in Chapter 6 of this volume.

Chapters 1–3 and 4–5 can be studied independently, and only Chapter 6 brings them together. Chapters 1–3 treat the Krichever–Novikov algebras and their applications to CFT. Chapters 4–5 treat Lax operator algebras and their applications to integrable systems. These two chapters can serve as an introduction to the theory of finite-dimensional integrable systems written from quite general point of view.

I am obliged to several people having made possible the appearance of this book, in the first place to my colleagues in the Steklov Mathematical Institute, especially the members of the Department of Geometry and Topology, and the head of the Department S.P.Novikov for the stimulating research atmosphere.

Two persons have played a special role in the appearance of this book and in all the preceding work: I. Krichever and M. Schlichenmaier. I have enjoyed discussions with I. Krichever since 1977, and in particular watching the emergence of the Krichever–Novikov algebras. In 2006 we were lucky to discover the Lax operator algebras together. Jointly with M. Schlichenmaier we obtained many results on the Krichever–Novikov algebras and their applications to the 2D CFT. I am grateful to the Universities of Mannheim and Luxembourg, where we were working together during my numerous visits. This was a great support for me, especially in the difficult 1990s. M. Schlichenmaier is now writing his own book on Krichever–Novikov algebras and generalizations [50]. I am sure that it will contain interesting topics which have been omitted here, such as deformation theory, more details on the multi-point algebras, classification of central extensions, etc.

Different parts of the investigation summarized in this book are supported by the Russian Foundation for Basic Research, by the Russian Academy of Sciences within the framework of the program "Fundamental problems of nonlinear dynamics", and by research projects of the Universities of Mannheim and Luxembourg.

Moscow, Oleg K. Sheinman
April 2012

Contents

Chapter 1

Krichever–Novikov algebras: basic definitions and structure theory

Let Σ be a compact Riemann surface of genus g with a given complex structure, or a smooth projective curve over $\mathbb{C}$, in terms of algebraic geometry. By marked points on Σ we mean an ordered set of mutually different points $P_1, \ldots, P_N \in \Sigma$, and P_∞, a distinguished point different from P_i for any i. Let

$$I = \{P_1, \ldots, P_N | N \geq 1\}, \quad A = I \cup \{P_\infty\}.$$

We will often use the notation Σ^* for $\Sigma \setminus A$, call Σ^* a punctured Riemann surface, and draw it something like the following:

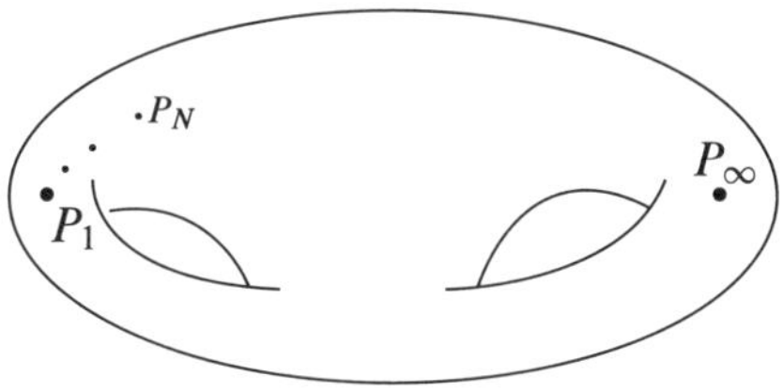

Figure 1.1. Riemann surface of genus 2 with marked points.

The idea of a Riemann surface with marked points in generic position originates in the string theory and two-dimensional conformal field theory (2D CFT). In the string theory, the points in I are called *incoming*, and P_∞ is called *outgoing*. The Riemann surface Σ is interpreted as a world sheet (diagram) of N interacting strings coming from infinity and producing one string going to infinity. It is assumed that in the course of time strings may divide into several contours which then may join again in various combinations. The trace of such an evolution is a (generically nontrivial) Riemann surface (Figure 1.2, page 3).

All results of this section are true for an arbitrary number of outgoing points. This case is considered in [44, 45, 50].

1.1 Current, vector field, and other Krichever–Novikov algebras

Let $\mathcal{A} := \mathcal{A}(\Sigma, I, P_\infty)$ be the associative algebra of functions on Σ regular except at the points $P \in A$ where they may have poles. Let $\mathfrak{g}$ be a finite-dimensional reductive

Lie algebra over $\mathbb{C}$. Then

$$\overline{\mathfrak{g}} = \mathfrak{g} \otimes_{\mathbb{C}} \mathcal{A} \tag{1.1.1}$$

is called a *current Krichever–Novikov algebra* [26, 57, 59, 61]. The bracket on $\overline{\mathfrak{g}}$ is given by the following relations:

$$[x \otimes A, y \otimes B] = [x, y] \otimes AB. \tag{1.1.2}$$

We will often omit the symbol $\otimes$ in our notation.

Let $\mathcal{V}$ denote the Lie algebra of meromorphic vector fields on Σ which also may have poles only at $P \in A$ [26–28]. We call them *Krichever–Novikov vector fields.*

For a Riemann sphere ($g = 0$) regarded as a completed complex line with the natural coordinate z, let us take $I = \{0\}$, $P_\infty = \infty$. Then $\mathcal{A}$ is the algebra of Laurent polynomials, $\overline{\mathfrak{g}}$ coincides with the conventional loop algebra [15], and $\mathcal{V}$ is the Witt algebra. Sometimes we call this case classical, for short.

The Lie algebra $\mathcal{V}$ acts by derivations on the elements of the algebra $\mathcal{A}$. It enables one to define the Lie algebra $\mathcal{D}^1$ of first order differential operators as a semidirect sum of $\mathcal{A}$ and $\mathcal{V}$. As a vector space $\mathcal{D}^1 = \mathcal{A} \oplus \mathcal{V}$. A Lie bracket is defined by means the following relation:

$$[A + e, B + f] = (eB - fA) + [e, f], \qquad A, B \in \mathcal{A}, \quad e, f \in \mathcal{V}. \tag{1.1.3}$$

Here, eA is a derivative of the function A along the vector field e. In local coordinates they write as $e = \tilde{e}\frac{d}{dz}$ and $eA = \tilde{e} \cdot \frac{dA}{dz}$.

Define also the Lie algebra $\mathcal{D}^1_{\mathfrak{g}}$ of differential operators related to $\overline{\mathfrak{g}}$ (which we call *Krichever–Novikov differential operators*). As a linear space we put $\mathcal{D}^1_{\mathfrak{g}} = \overline{\mathfrak{g}} \oplus \mathcal{V}$. A Lie bracket is given by means of the brackets on $\overline{\mathfrak{g}}$ and $\mathcal{V}$, and additional requirements

$$[e, x \otimes A] = -[x \otimes A, e] = x \otimes (eA). \tag{1.1.4}$$

In particular, for $\mathfrak{g} = \mathfrak{gl}(1)$ we obtain $\mathcal{D}^1_{\mathfrak{g}} = \mathcal{D}^1$.

1.2 Meromorphic λ-forms and Krichever–Novikov duality

Let $\mathcal{K}$ denote the canonical bundle on Σ. The sheaf of its local sections is a sheaf of 1-forms (*differentials*) on Σ. Following the established practice, we will not distinguish between a linear bundle and the sheaf of its local sections. For every $\lambda \in \mathbb{Z}$ let us consider the bundle $\mathcal{K}^\lambda = \mathcal{K}^{\otimes\lambda}$. It is conventional that $\mathcal{K}^0 = \mathcal{O}$ is a trivial bundle, and $\mathcal{K}^{-1} = \mathcal{K}^*$ is the (holomorphic, linear) tangent bundle. Indeed, having fixed theta characteristics, i.e., a bundle S such that $S^{\otimes 2} = \mathcal{K}$, we could consider $\lambda \in \frac{1}{2}\mathbb{Z}$. Denote by $\mathcal{F}^\lambda$ the infinite-dimensional vector space of global meromorphic sections of the bundle $\mathcal{K}^\lambda$ holomorphic on $\Sigma \setminus A$. The elements of the space $\mathcal{F}^\lambda$ are called

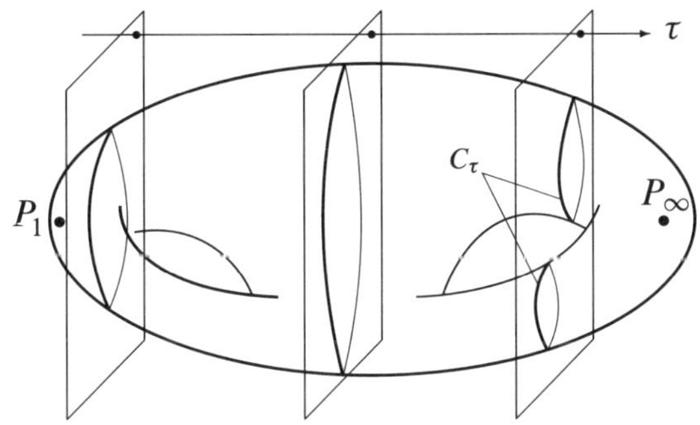

Figure 1.2. Worldsheet of a string with one incoming, and one outgoing point

(meromorphic) tensors of weight λ. For $\lambda \geq 0$, the meromorpic tensors of weight λ are called *λ-forms*, or *forms of weight λ*.

The following cases are of special interest: functions ($\lambda = 0$), vector fields ($\lambda = -1$), 1-forms ($\lambda = 1$) and quadratic differentials ($\lambda = 2$). Above, the space of functions is denoted by $\mathcal{A}$, and the space of vector fields by $\mathcal{V}$.

Multiplying sections by functions, we obtain sections again. In this way the spaces $\mathcal{F}^\lambda$ become $\mathcal{A}$-modules. The action of vector fields on forms by Lie derivative turns the $\mathcal{F}^\lambda$ into $\mathcal{V}$-modules. In local coordinates the Lie derivative is given by

$$L_e\omega = \left(\tilde{e}(z)\frac{dw}{dz}(z) + \lambda\, w(z)\frac{d\tilde{e}}{dz}(z)\right) dz^\lambda, \tag{1.2.1}$$

where $e = \tilde{e}(z)\frac{d}{dz} \in \mathcal{V}$, $\omega = w(z)\, dz^\lambda \in \mathcal{F}^\lambda$.

Combining these definitions we equip $\mathcal{F}^\lambda$ with the structure of $\mathcal{D}^1$-modules: $(A+e)\omega = A\cdot\omega + L_e\omega$ where $A \in \mathcal{A}$, $e \in \mathcal{V}$ è $\omega \in \mathcal{F}^\lambda$. Universal constructions enable one to consider the algebras of differential operators of an arbitrary order [45,47].

There is a unique meromorphic differential ρ on Σ holomorphic outside A, having simple poles at the points in A, given positive residues at the points in I, given negative residue at P_∞ (such that $\sum_{P\in I} \operatorname{res}_P(\rho) + \operatorname{res}_{P_\infty}(\rho) = 0$), and pure imaginary periods [26,41].

For a fixed $R \in \Sigma \setminus A$ the $u(P) = \operatorname{Re}\int_R^P \rho$ is an unambiguously defined harmonic function. A family of level lines $C_\tau = \{P \in \Sigma \mid u(P) = \tau\}$, $\tau \in \mathbb{R}$ defines a fibration on $\Sigma \setminus A$. Every curve C_τ isolates points in I from P_∞. For every $\tau \ll 0$ ($\tau \gg 0$) the level line C_τ is a disjoint union of deformed circles C_i around points P_i, $i = 1, \ldots, N$ (deformed circle C_∞ around P_∞). We call every such level line, or every cycle homological to such level line, a *separating cycle* and denote it C_S. For example, all plane sections in Figure 1.2 are separating cycles.

In the framework of the above mentioned interpretation of Σ as the world sheet of a string, we interpret τ as the time, then we can think of the contour C_τ as of the string at the instant τ [26–28] (see Figure 1.2). The vertical planes in Figure 1.2 are only needed to symbolize that contours are attributed to certain time points.

Definition 1.1. *Krichever–Novikov duality* is given by pairing of the spaces $\mathcal{F}^{\lambda}$ and $\mathcal{F}^{1-\lambda}$ by means of the inner product

$$\mathcal{F}^{\lambda} \times \mathcal{F}^{1-\lambda} \to \mathbb{C},$$
$$\langle f, g\rangle := \frac{1}{2\pi \mathrm{i}} \int_{C_S} fg = \sum_{P\in I} \mathrm{res}_P(fg) = -\mathrm{res}_{P_\infty}(fg), \tag{1.2.2}$$

where C_S is an arbitrary separating cocycle.

The last relation is due to the residue theorem. Observe that the integral in (1.2.2) does not depend on the choice of a separating cocycle. It follows from the existence of dual bases constructed in the next section that the inner product (1.2.2) is nondegenerate.

1.3 Krichever–Novikov bases

Krichever and Novikov [26] introduced certain bases (*Krichever–Novikov bases*) in the spaces $\mathcal{F}^{\lambda}$ on Riemann surfaces with two marked points. For $g = 0$ the Krichever–Novikov bases coincide with conventional Fourier–Laurent bases. A multipoint generalization is given in [44, 45] (see also [40] for some results in this direction). Considerations related to Krichever–Novikov bases heavily use the Riemann–Roch theorem and seem to be the easiest of all in the case of nonspecial divisors (we mean divisors supported at $A \subset \Sigma$ here). As it usually happens, two of the three most important types of tensors (functions, 1-forms and vector fields) are related to special divisors. This also depends on g and N. For this reason it is easier to define Krichever–Novikov bases for a generic case and for most important special cases rather than try to give a general definition.

Following [44, 45], let us define Krichever–Novikov bases for the tensors of an arbitrary weight λ on a Riemann surface with N marked points in a generic position. For a fixed λ, we associate a basis element to every pair of an arbitrary integer and an arbitrary incoming point. Thus for arbitrary $n \in \mathbb{Z}$, $p = 1, \dots, N$ we produce a basis element $f^{\lambda}_{n,p} \in \mathcal{F}^{\lambda}$. We do it in such way that the following duality relations are satisfied with respect to the pairing (1.2.2):

$$\langle f^{\lambda}_{n,p}, f^{1-\lambda}_{m,r}\rangle = \delta^{m}_{-n} \cdot \delta^{r}_{p}. \tag{1.3.1}$$

In particular, this implies that the pairing (1.2.2) is nondegenerate.

By definition, at the incoming points the basis elements must satisfy

$$\mathrm{ord}_{P_i}\left(f^{\lambda}_{n,p}\right) = (n + 1 - \lambda) - \delta^{p}_{i}, \quad i = 1, \dots, N. \tag{1.3.2}$$

At the outgoing point P_∞, for $g \geq 2$, $\lambda \neq 0, 1$ we require that

$$\operatorname{ord}_{P_\infty}\left(f^\lambda_{n,p}\right) = -N \cdot (n + 1 - \lambda) + (2\lambda - 1)(g - 1). \tag{1.3.3}$$

For the points in A in a generic position the space of the elements satisfying (1.3.2), (1.3.3) is one-dimensional by the Riemann–Roch theorem [42]. To normalize the basis element we choose local coordinates z_i at the points P_i, and require

$$f^\lambda_{n,p}(z_p) = z_p^{n-\lambda}\left(1 + O(z_p)\right)\left(dz_p\right)^\lambda, \quad p = 1, \ldots, N. \tag{1.3.4}$$

Observe that only the first jet of the local coordinate is of importance for the normalization. For $g = 0$ no genericity condition is imposed on A. The element $f^\lambda_{n,p}$ can be uniquely defined for all combinations of values g and λ. For the necessary modifications see [44, 45, 53].

There are explicit descriptions of basis elements $f^\lambda_{n,p}$ in terms of rational functions for $g = 0$, Weierstrass σ-function for $g = 1$, theta-functions and prime forms for $g \geq 1$ [43]. For $g = 0$ and $g = 1$ such descriptions can be found also in [**?**, Sects. 2 and 7]. For the descriptions in terms of the Weierstrass $\wp$-function see [39, 46]. In particular, it can be observed from the explicit representations that the basis elements analytically depend on the complex structure (moduli) of the Riemann surface.

In the following cases we use a special notation:

$$A_{n,p} = f^0_{n,p}, \quad e_{n,p} = f^{-1}_{n,p}, \quad \omega^{n,p} = f^1_{-n,p}, \quad \Omega^{n,p} = f^2_{-n,p}\,. \tag{1.3.5}$$

For $g \geq 1$ and $N = 1$ this coincides, up to a shift of n, with the Krichever–Novikov notation in [26–28].

Let us describe, following [26], the deviation from the prescriptions (1.3.2)–(1.3.4) in the most important cases, i.e., for functions ($\lambda = 0$) and 1-forms ($\lambda = 1$), for two marked points ($N = 1$), and $g \geq 2$.

It is conventional to denote the marked points by $P_\pm$, an $A_{n,1}$ by A_n, and an $\omega^{n,1}$ by ω^n in the case of two marked points. Let $z_\pm$ be local coordinates at $P_\pm$, respectively. The prescription for A_n and ω^n distinguishes from the generic case only for $-g \leq n \leq 0$. Thus the complete prescription for A_n is written as

$$\begin{aligned}
&A_n(z_+) = z_+^n\left(1 + \mathcal{O}(z_+)\right),\\
&A_n(z_-) = \begin{cases} \mathcal{O}(z_-^{-n-g}), & \text{for } n < -g \text{ and } n > 0,\\ \mathcal{O}(z_-^{-n-g-1}) & \text{for } -g \leq n \leq -1.\end{cases}\\
&A_0 \equiv 1,
\end{aligned} \tag{1.3.6}$$

and for ω^n

$$\begin{aligned} \omega^n(z_+) &= z_+^{-n-1}\big(1+\mathcal{O}(z_+)\big)dz_+,\\ \omega^n(z_-) &= \begin{cases} \mathcal{O}(z_-^{n+g-1})dz_- & \text{for } n<-g \text{ and } n>0,\\ \mathcal{O}(z_-^{n+g})dz_- & \text{for } -g\le n\le -1, \end{cases}\\ \omega^0(z_\pm) &= \pm\frac{dz_\pm}{z_\pm}\big(1+\mathcal{O}(z_\pm)\big), \end{aligned} \tag{1.3.7}$$

In this notation, A_n and ω^n are dual elements.

It is an easy exercise to prove that for $g = 0$ (Σ being a Riemann sphere), and a special choice of marked points the Krichever–Novikov function algebra is just the algebra of Laurent polynomials:

$$\mathcal{A} = \mathbb{C}[z,z^{-1}] \quad \text{for } g=0,\ P_+=0,\ P_-=\infty.$$

We have $A_n = z^n$ in this case.

It is a little more difficult to prove that for $g = 1$ (elliptic curve), $P_\pm = \pm z_0$ (z_0 being a generic point of the curve)

$$\begin{aligned} A_n &= \frac{\sigma^n(z-z_0)\sigma(z+(2n+1)z_0)}{\sigma^{n+1}(z+z_0)\sigma((2n+2)z_0)}\sigma^{n+1}(2z_0), \quad n\ne 0,-1,\\ A_{-1}(z) &= \frac{\sigma^2(z)\sigma(2z_0)}{\sigma(z+z_0)\sigma(z-z_0)\sigma^2(z_0)}, \qquad A_0\equiv 1 \end{aligned} \tag{1.3.8}$$

where σ denotes the Weierstrass σ-function (see [26] for the proof). The expression for a basis function in the many point case is given in Section 3.6.

For completeness, let us give here the formulae for the basis elements $e_m \in \mathcal{V}$, following from (1.3.2)–(1.3.4) ($\lambda = -1$), in the two-point case ($N = 1$):

$$\begin{aligned} e_n(z_+) &= z_+^{n+1}\big(1+\mathcal{O}(z_+)\big)\frac{\partial}{\partial z_+},\\ e_n(z_-) &= \varepsilon z_-^{-n+1-3g}\big(1+\mathcal{O}(z_-)\big)\frac{\partial}{\partial z_-}. \end{aligned} \tag{1.3.9}$$

1.4 Almost-graded structure, triangle decompositions

For $g = 0$, $N = 1$ the Lie algebras defined in Section 1.1 are graded. These are just conventional loop algebra, Virasoro algebra, and their combination first order differential operator algebra. Their grading is a necessary tool of their structure theory and highest weight representation theory. For $g \ge 1$, or $g = 0$ and $N > 1$, there is no grading. It is a fundamental observation of Krichever and Novikov [26–28] that a weaker property, namely almost-grading, is already sufficient for developing the structure theory and representation theory in these more general cases.

Let $\mathcal{V}$ denote an arbitrary associative or Lie algebra in this section. $\mathcal{V}$ is called *almost graded* if a decomposition into a direct sum of vector spaces is given: $\mathcal{V} = \bigoplus_{n\in\mathbb{Z}} \mathcal{V}_n$, where $\dim \mathcal{V}_n < \infty$, and constants R and S exist such that

$$\mathcal{V}_n \cdot \mathcal{V}_m \subseteq \bigoplus_{h=n+m-R}^{n+m+S} \mathcal{V}_h, \qquad \forall n,m \in \mathbb{Z} \tag{1.4.1}$$

where $\cdot$ denotes the multiplication in $\mathcal{V}$. The elements of $\mathcal{V}_n$ are called *homogeneous of degree n*. Let $\mathcal{V} = \oplus_{n\in\mathbb{Z}}\mathcal{V}_n$ be an almost graded algebra and M be a $\mathcal{V}$-module. M is called *almost graded* if a decomposition into a direct sum of vector spaces is given: $M = \bigoplus_{m\in\mathbb{Z}} M_m$, where $\dim M_m < \infty$ and there are constants T and U such that

$$\mathcal{V}_n M_m \subseteq \bigoplus_{h=n+m-T}^{n+m+U} M_h, \qquad \forall n,m \in \mathbb{Z}\,. \tag{1.4.2}$$

The elements of M_n are called *homogeneous of degree n*. We call the numbers $R+S$, $T+U$ *diversities* of gradings. In particular, zero diversity corresponds to a graded algebra.

Let us define $\mathcal{F}_n^\lambda \subset \mathcal{F}^\lambda$ as a subspace generated by all elements $f_{n,p}^\lambda$, $p = 1,\dots,N$. In analogy with (1.3.5)

$$\mathcal{A}_n = \mathcal{F}^{\lambda^0}_{\ n}, \quad \mathcal{V}_n = \mathcal{F}^{\lambda^{-1}}_{\ n}\,. \tag{1.4.3}$$

Proposition 1.2 ([44, 45]). *The following direct decompositions hold:*

$$\mathcal{A} = \bigoplus_{n\in\mathbb{Z}} \mathcal{A}_n, \quad \mathcal{V} = \bigoplus_{n\in\mathbb{Z}} \mathcal{V}_n, \quad \mathcal{F}^\lambda = \bigoplus_{n\in\mathbb{Z}} \mathcal{F}_n^\lambda\,.$$

They give the structures of an almost-graded algebra on $\mathcal{A}$, an almost-graded Lie algebras on $\mathcal{V}$ and $\mathcal{D}^1$, and an almost-graded module over them on $\mathcal{F}^\lambda$. In particular constants $K, L \in \mathbb{N}$ exist such that for all $n, m \in \mathbb{Z}$

$$A_{n,p} \cdot A_{m,r} = \delta_p^r\, A_{n+m,p} + \sum_{h=n+m+1}^{n+m+K} \sum_{s=1}^{N} \alpha_{(n,p),(m,r)}^{(h,s)} A_{h,s},$$

$$[e_{n,p}, e_{m,r}] = \delta_p^r\,(m-n)\, e_{n+m,p} + \sum_{h=n+m+1}^{n+m+L} \sum_{s=1}^{N} \gamma_{(n,p),(m,r)}^{(h,s)} e_{h,s},$$

with suitable coefficients $\alpha_{(n,p),(m,r)}^{(h,s)}, \gamma_{(n,p),(m,r)}^{(h,s)} \in \mathbb{C}$.

As a vector space, $\mathcal{A}$ can be decomposed as follows:

$$\mathcal{A} = \mathcal{A}_+ \oplus \mathcal{A}_{(0)} \oplus \mathcal{A}_-, \tag{1.4.4}$$

where

$$\begin{aligned} \mathcal{A}_+ &:= \langle A_{n,p} \mid n \geq 1, p = 1, \ldots, N \rangle, \\ \mathcal{A}_- &:= \langle A_{n,p} \mid n \leq -K - 1, p = 1, \ldots, N \rangle, \\ \mathcal{A}_{(0)} &:= \langle A_{n,p} \mid -K \leq n \leq 0, p = 1, \ldots, N \rangle, \end{aligned} \tag{1.4.5}$$

and $\mathcal{V}$ can be decomposed as

$$\mathcal{V} = \mathcal{V}_+ \oplus \mathcal{V}_{(0)} \oplus \mathcal{V}_-, \tag{1.4.6}$$

where

$$\begin{aligned} \mathcal{V}_+ &:= \langle e_{n,p} \mid n \geq 1, p = 1, \ldots, N \rangle, \\ \mathcal{V}_- &:= \langle e_{n,p} \mid n \leq -L - 1, p = 1, \ldots, N \rangle\,, \\ \mathcal{V}_{(0)} &:= \langle e_{n,p} \mid -L \leq n \leq 0, p = 1, \ldots, N \rangle\,. \end{aligned} \tag{1.4.7}$$

We call the relations (1.4.4), (1.4.7) *triangle decompositions*. In a similar way we obtain a triangle decomposition of the Lie algebra $\mathcal{D}^1$.

Due to almost grading, the subspaces $\mathcal{A}_\pm$ and $\mathcal{V}_\pm$ are subalgebras but $\mathcal{A}_{(0)}$ and $\mathcal{V}_{(0)}$ do not appear as such in general. We use the term *critical strip* to denote them.

Observe that $\mathcal{A}_+$ and $\mathcal{V}_+$ can be characterized as algebras of functions having zeroes of at least order one (vector fields having zeroes of at least order two, respectively) at P_i, $i = 1, \ldots, N$. These subalgebras are not maximal. They can be extended by considering all elements regular at the points P_i. It can be done by adding the set $\{A_{0,p}, p = 1, \ldots, N\}$ ($\{e_{0,p}, e_{-1,p}, p = 1, \ldots, N\}$, respectively) to their bases.

In turn, $\mathcal{A}_-$ and $\mathcal{V}_-$ can be extended in such way that they contained all elements regular at P_∞. It is explained in [53] in detail. In a similar way, for every nonnegative $p \in \mathbb{N}$, let $\mathcal{V}_-^{(p)}$ be a subalgebra of vector fields of the order of at least $p + 1$ at P_∞, and $\mathcal{A}_-^{(p)}$ be a subalgebra of functions of the order of at least p at P_∞. Then we obtain the decompositions

$$\mathcal{V} = \mathcal{V}_+ \oplus \mathcal{V}_{(0)}^{(p)} \oplus \mathcal{V}_-^{(p)}, \quad \mathcal{A} = \mathcal{A}_+ \oplus \mathcal{A}_{(0)}^{(p)} \oplus \mathcal{A}_-^{(p)} \tag{1.4.8}$$

with "critical strips" $\mathcal{V}_{(0)}^{(p)}$ and $\mathcal{A}_{(0)}^{(p)}$ which possess only the structure of subspaces. Such triangle decompositions arise in the Chapter 3 below, for example the space $\mathcal{V}_{(0)}^{(1)}$ which we call a *reduced critical strip* is of special interest. For $g \geq 2$ its dimension is equal to

$$\dim \mathcal{V}_{(0)}^{(1)} = N + N + (3g - 3) + 1 + 1 = 2N + 3g - 1. \tag{1.4.9}$$

Here, the first two summands correspond to the dimensions of $\mathcal{V}_0$ and $\mathcal{V}_{-1}$. The middle term comes from the basis vector fields having poles at $P_i, i = 1, \dots, N$, and at P_∞, as well. The sum $1 + 1$ corresponds to the basis vector fields of exact orders 0 and 1, respectively, at P_∞.

It is easy to continue the almost-grading to the current algebra $\overline{\mathfrak{g}}$ by setting $\deg(x \otimes A_{n,p}) = n$. As above, we obtain a triangular decomposition

$$\overline{\mathfrak{g}} = \overline{\mathfrak{g}}_+ \oplus \overline{\mathfrak{g}}_{(0)} \oplus \overline{\mathfrak{g}}_-, \quad \text{where} \quad \overline{\mathfrak{g}}_\beta = \mathfrak{g} \otimes \mathcal{A}_\beta, \quad \beta \in \{+, (0), -\} \tag{1.4.10}$$

In particular, $\overline{\mathfrak{g}}_\pm$ are subalgebras. As above, they can be extended by means adding certain elements of zero degree.

The finite-dimensional Lie algebra $\mathfrak{g}$ is naturally embedded into $\overline{\mathfrak{g}}$, more precisely, into the subspace $\overline{\mathfrak{g}}_0$. For two marked points ($N = 1$) this is simply the equality $A_0 \equiv 1$ (1.3.6). In the case $N > 1$ it is implied by the following simple lemma.

Lemma 1.3 ([52]).

$$1 = \sum_{p=1}^{N} A_{0,p}. \tag{1.4.11}$$

Proof. By the duality relation (1.3.1) we have

$$1 = \sum_{n \in \mathbb{Z}} \sum_{p=1}^{N} \langle 1, \omega^{n,p} \rangle A_{n,p}.$$

By (1.2.2)

$$\langle 1, \omega^{n,p} \rangle = \sum_{P \in I} \operatorname{res}_P \omega^{n,p} = -\operatorname{res}_{P_\infty} \omega^{n,p}.$$

By (1.3.2), $\operatorname{ord}_{P_i} \omega^{n,p} = -n - \delta_p^i \geq 0$ for $n < 0$, and by (1.3.3), $\operatorname{ord}_{P_\infty} \omega^{n,p} = Nn + g - 1 \geq 0$ for $n > 0$. Hence

$$\langle 1, \omega^{n,p} \rangle = \begin{cases} 0, & n \neq 0; \\ 1, & n = 0. \end{cases} \tag{1.4.12}$$

The second case follows by (1.3.4). □

1.5 Central extensions and 2-cohomology; Virasoro-type algebras

In this section, let $\mathcal{V}$ denote an arbitrary Lie algebra. A central extension of $\mathcal{V}$, by definition, is a short exact sequence of Lie algebras

$$0 \longrightarrow \mathbb{C} \xrightarrow{\ i\ } \widehat{\mathcal{V}} \xrightarrow{\ p\ } \mathcal{V} \longrightarrow 0 \tag{1.5.1}$$

where $\mathrm{Im}(i) = \ker(p)$ is the center of $\widehat{\mathcal{V}}$. The Lie algebra $\widehat{\mathcal{V}}$ is often called a central extension of $\mathcal{V}$ as well.

Two central extensions $\widehat{\mathcal{V}}$ and $\widehat{\mathcal{V}}'$ are called equivalent if there exists an isomorphism e (*equivalence*) such that the following diagram is commutative:

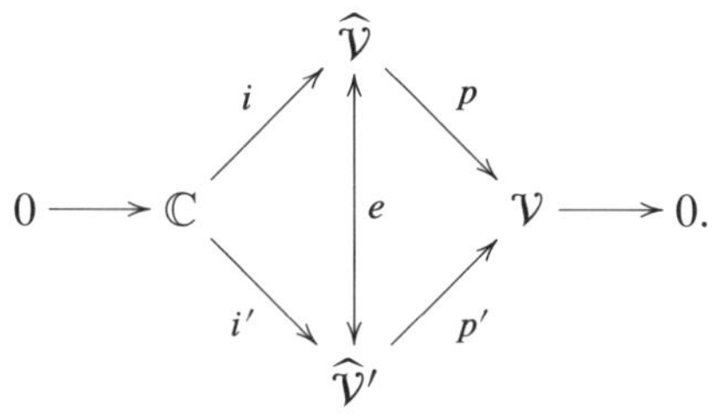

(1.5.2)

A *2-cocycle* on $\mathcal{V}$ is a skew-symmetric bilinear form γ satisfying the relation

$$\gamma([f,g],h) + \gamma([g,h],f) + \gamma([h,f],g) = 0, \quad \forall f,g,h \in \mathcal{V}. \tag{1.5.3}$$

If $\gamma(f,g) = \phi([f,g])$ where $\phi \in \mathcal{V}^*$ then γ is called a *coboundary* (and denoted by $\delta\phi$). If $\gamma - \gamma' = \delta\phi$ then γ and γ' are called *cohomological*.

Given a 2-cocycle γ on $\mathcal{V}$ one may define a central extension by setting $\widehat{\mathcal{V}} = \mathcal{V} \oplus \mathbb{C}t$ and

$$[f,g] = [f,g]_{\mathcal{V}} + \gamma(f,g)\cdot t, \quad [t,\mathcal{V}] = 0, \tag{1.5.4}$$

where $f,g \in \mathcal{V}$, t is a formal generator of the 1-dimensional space $\mathbb{C}t$, $[\cdot,\cdot]$ denotes a commutator to be defined in $\widehat{\mathcal{V}}$, and $[\cdot,\cdot]_{\mathcal{V}}$ is a given bracket in $\mathcal{V}$. Since $\mathcal{V}$ is embedded into $\widehat{\mathcal{V}}$, there is no difference between these brackets on $\mathcal{V}$. If $\mathcal{V}$ has only trivial center, the center of $\widehat{\mathcal{V}}$ is equal to $\mathbb{C}t$. A central extension $\widehat{\mathcal{V}}$ can be obtained in this way if there exists a section $s : \mathcal{V} \to \widehat{\mathcal{V}}$ such that s is an embedding of Lie algebras as well. In this case one can construct γ by letting $[f,g] - s([f,g]_{\mathcal{V}}) = \gamma(f,g)t$. Two central extensions are equivalent if and only if their defining cocycles are cohomological. Thus the space of classes of equivalent central extensions is in a one-to-one correspondence with $\mathrm{H}^2(\mathcal{V},\mathbb{C})$.

Example 1.4. (1) $\mathcal{V} = \mathbb{C}^{2n}$ with a structure of a commutative Lie algebra, γ is a symplectic form on $\mathcal{V}$. Then $\widehat{\mathcal{V}}$ is Heisenberg algebra.

(2) For $\mathcal{V} = \overline{\mathfrak{gl}(n)}$ (a current algebra of Section 1.1) a conventional cocycle is given by $\gamma(f,g) = \mathrm{res}_{P_+} \mathrm{tr}(f \cdot dg)$.

Definition 1.5. Let a decomposition $\mathcal{V} = \bigoplus_{n\in\mathbb{Z}} \mathcal{V}_n$ give an almost graded structure. A cocycle γ on $\mathcal{V}$ is called local (with respect to the almost graded structure) if there exist $M_1, M_2 \in \mathbb{Z}$ such that

$$\forall n,m \in \mathbb{Z}: \quad \gamma(\mathcal{V}_n,\mathcal{V}_m) \neq 0 \implies M_2 \le n+m \le M_1. \tag{1.5.5}$$

The constants M_1 and M_2 are called the upper and the lower boundaries, respectively, of the cocycle γ.

Provided $\deg(t) = 0$ a central extension $\widehat{\mathcal{V}}$ is almost graded if and only if it is given by a local cocycle γ. For this reason almost-graded central extensions may occur by the name of *local central extensions*. The almost-graded central extensions are in a one-to-one correspondence with the 2-cohomology classes which can be represented by local cocycles. Observe that it is not the case that every representative of such a class is a local cocycle. Moreover, a local representative is, basically, unique.

Below, we consider cocycles of geometric origin. First, we do it for $\mathcal{A}$, $\mathcal{V}$, and $\mathcal{D}^1$. Complete presentation for those cases is given in [48, 50]. We refer to these works for proofs of the statements in the remainder of the section, and for further details.

For the $\mathcal{A}$ regarded as a commutative Lie algebra, each skew-symmetric bilinear form is a cocycle. Let C be an arbitrary (not necessary connected) smooth contour in Σ^* (the punctured Riemann surface), then the bilinear form defined by the relations

$$\gamma_C^{(f)} : \mathcal{A} \times \mathcal{A} \to \mathbb{C}, \quad \gamma_C^{(f)}(g,h) := \frac{1}{2\pi \mathrm{i}} \int_C g\,dh \tag{1.5.6}$$

is skew-symmetric, hence a cocycle. Observe that any (smooth) contour homological to C gives the same cocycle. This cocycle is $\mathcal{V}$-*invariant*, i.e.,

$$\gamma_C^{(f)}(eg,h) = \gamma_C^{(f)}(eh,g), \quad \forall e \in \mathcal{V}, \ \forall g,h \in \mathcal{A}. \tag{1.5.7}$$

For the Lie algebra $\mathcal{V}$ of vector fields, following [26], let us generalize the Virasoro–Gelfand–Fuchs cocycle. To this end let us first choose a projective connection. *A projective connection* is, by definition, a function R of a point and local coordinates on the Riemann surface transforming as follows under the transformation of coordinates $u = u(z)$:

$$R(u)u_z^2 = R(z) + \frac{u_{zzz}}{u_z} - \frac{3}{2}\left(\frac{u_{zz}}{u_z}\right)^2.$$

Let R be a holomorphic projective connection, C be an arbitrary cycle. Associate a cocycle with the pair (C, R) by means of the following relation:

$$\gamma_{C,R}^{(v)}(e,f) := \frac{1}{24\pi\,\mathrm{i}} \int_C \left(\frac{1}{2}(\tilde{e}'''\tilde{f} - \tilde{e}\tilde{f}''') - R\cdot\left(\tilde{e}'\tilde{f} - \tilde{e}\tilde{f}'\right)\right) dz. \tag{1.5.8}$$

Here, $e = \tilde{e}\frac{d}{dz}$, $f = \tilde{f}\frac{d}{dz}$, $\tilde{e}$, and $\tilde{f}$ are local meromorphic functions. Due to the transformation law of R, the integrand is a 1-form on the Riemann surface, i.e., the integral is well-defined. Another choice of a holomorphic projective connection results in a cohomological cocycle, hence an equivalent central extension. It is just as true if we choose among meromorphic projective connections having poles at the points of A.

The cocycles of the two types just considered can be continued to the whole Lie algebra $\mathcal{D}^1$, provided they are set to zero on the subspaces complementary to their domains of definition, and on mixed pairs (f, e), $f \in \mathcal{A}$, $e \in \mathcal{V}$. Moreover, every mixing-free cocycle on $\mathcal{V}$, and every mixing-free $\mathcal{V}$-invariant cocycle on $\mathcal{A}$ can be continued on $\mathcal{D}^1$.

There is one more kind of cocycle which does not vanish on mixed pairs. We call them *mixing cocycles*. To define a mixing cocycle, let us choose a meromorphic affine connection T holomorphic outside the marked points. An *affine connection* [45, 48, 63, 64] is a function T of a point and local coordinates on the Riemann surface that transforms as follows under a transformation $u = u(z)$ of the coordinates:

$$T(u)u_z = T(z) + \frac{u_{zz}}{u_z}.$$

Then

$$\gamma^{(m)}_{C,T}(e, g) = -\gamma^{(m)}_{C,T}(g, e) = \frac{1}{2\pi \mathrm{i}} \int_C \left(\tilde{e} \cdot g'' + T \cdot (\tilde{e} \cdot g')\right) dz \qquad (1.5.9)$$

is a well-defined 2-cocycle. As above, its cohomology class does not depend on the choice of the affine connection.

Let us consider cocycles obtained by integration over the separating cycle C_S. We call such cocycles *geometrical*, or *standard*. We write γ_S instead γ_{C_S}. Obviously, these cocycles can be expressed via residues at the points of I or, equivalently, at P_∞.

Proposition 1.6 ([26, 45]). *The cocycles* (1.5.6), (1.5.8), (1.5.9) *given by integration over a separating cycle are local, with a zero upper boundary.*

The proof is obtained by considering orders of the integrands at the marked points.

The cocycles remain local when replacing R or T with another meromorphic connections having poles only at the marked points.

By locality, the geometrical cocycles give almost-graded central extensions $\widehat{\mathcal{A}}$, $\widehat{\mathcal{V}}$, and $\widehat{\mathcal{D}}^1$. In particular, $\widehat{\mathcal{V}}$ is a generalization of the Virasoro algebra onto the case of Riemann surfaces of positive genus with marked points (a conventional Virasoro algebra is produced in the case of Riemann sphere with the marked points 0 and ∞).

Definition 1.7. A central extension $\widehat{\mathcal{V}}$ of the Lie algebra $\mathcal{V}$, given by a cocycle of the form (1.5.8) where $C = C_S$ is a separating cycle, is called a *Virasoro-type Krichever–Novikov algebra* .

By removing the geometrical cocycles (at least for a holomorphic R) on the subalgebras $\mathcal{A}_\pm$ and $\mathcal{V}_\pm$, the latter are naturally identified with the subalgebras $\widehat{\mathcal{A}}_\pm$ and $\widehat{\mathcal{V}}_\pm$ in $\widehat{\mathcal{A}}$ and $\widehat{\mathcal{V}}$, respectively.

The following theorem is one of the main results of [48].

Theorem 1.8.

(a) *Every local cocycle on $\mathcal{A}$ satisfying the $\mathcal{V}$-invariance condition is a multiple (over $\mathbb{C}$) of $\gamma_S^{(f)}$. The cohomology class of $\gamma_S^{(f)}$ is nontrivial.*

(b) *Every local cocycle on $\mathcal{V}$ is cohomological to a multiple of $\gamma^{(v)}_{S,R}$. The cohomology class of $\gamma^{(v)}_{S,R}$ is nontrivial. Every comologically nontrivial local cocycle is a multiple of $\gamma^{(v)}_{C,R}(e, f)$ for some meromorphic projective connection R holomorphic outside A.*

(c) *Every local cocycle on $\mathcal{D}^1$, up to a coboundary, is a linear combination of the cocycles $\gamma^{(f)}_S$, $\gamma^{(v)}_{S,R}$ and $\gamma^{(m)}_{S,T}$, i.e.,*

$$\gamma = r_1\gamma^{(f)}_S + r_2\gamma^{(m)}_{S,T} + r_3\gamma^{(v)}_{S,R} + \textit{coboundary}, \quad r_1, r_2, r_3 \in \mathbb{C}. \tag{1.5.10}$$

The cohomology classes of the three cocycles above are linear independent. If $r_2 \neq 0$ and $r_3 \neq 0$, then there exist meromorphic R' and T' holomorphic outside A such that $\gamma = r_1\gamma^{(f)}_S + r_2\gamma^{(m)}_{S,T'} + r_3\gamma^{(v)}_{S,R'}$.

1.6 Affine Krichever–Novikov, in particular Kac–Moody, algebras

Let $\mathfrak{g}$ be a finite-dimennsional reductive Lie algebra. We have introduced an almost-graded current algebra $\overline{\mathfrak{g}}$. Here we will investigate its central extensions. Among those, the most important are affine Krichever–Novikov algebras.

Let α be a symmetric invariant bilinear form where invariance means that $\alpha([x, y], z) = \alpha(x, [y, z])$ for any triple of arguments.

Definition 1.9. An *affine Krichever–Novikov algebra* is a vector space of the form $\widehat{\mathfrak{g}} = \overline{\mathfrak{g}} \oplus \mathbb{C}t$ (t is a formal central generator) equipped with a Lie bracket

$$[x \otimes f, y \otimes g] = [x, y] \otimes (fg) + \gamma(x \otimes f, y \otimes g) \cdot t, \quad [t, \widehat{\mathfrak{g}}] = 0, \tag{1.6.1}$$

where

$$\gamma(x \otimes f, y \otimes g) = \frac{\alpha(x, y)}{2\pi i} \int_{C_S} f dg \tag{1.6.2}$$

is a cocycle on the function algebra given by integration over a separating cocycle C_S.

By definition, $\widehat{\mathfrak{g}}$ is a central extension of the Lie algebra $\overline{\mathfrak{g}}$. In the particular case of the Riemann sphere with marked points 0 and ∞, and a semisimple Lie algebra $\mathfrak{g}$, Definition 1.9 gives an affine Kac–Moody algebra. The extension $\widehat{\mathfrak{g}}$ depends on the choice of the form α. If we want this fact to be reflected in our notation, we write $\widehat{\mathfrak{g}}_{\alpha,S}$.

As a central extension, the affine Krichever–Novikov algebra is given by a local cocycle γ_S. Hence we are able to continue the almost-grading on $\overline{\mathfrak{g}}$ onto the whole $\widehat{\mathfrak{g}}$ by setting $\deg t = 0$ and $\deg(x \otimes A_{n,p}) = n$. As earlier, we obtain a triangle decomposition

$$\widehat{\mathfrak{g}}_{\alpha,S} = \widehat{\mathfrak{g}}_+ \oplus \widehat{\mathfrak{g}}_{(0)} \oplus \widehat{\mathfrak{g}}_- \quad \text{with} \quad \widehat{\mathfrak{g}}_\pm \cong \overline{\mathfrak{g}}_\pm \quad \text{and} \quad \widehat{\mathfrak{g}}_{(0)} = \overline{\mathfrak{g}}_{(0)} \oplus \mathbb{C}\cdot t\,. \qquad (1.6.3)$$

Instead of integration over a separating cycle in (1.6.2) we could integrate over another cycle C, and obtain another one, not equivalent and not isomorphic in general, central extension $\widehat{\mathfrak{g}}_{\alpha,C}$. Moreover, there is no a priori reason for an arbitrary cocycle on $\overline{\mathfrak{g}}$ to be of that type, i.e., to be given by an invariant symmetric bilinear form α and contour integration of the 1-form $f dg$.

Before we formulate the corresponding results, let us extend the definition (1.5.7) of $\mathcal{V}$-invariance so that it worked for the cocycles on $\overline{\mathfrak{g}}$.

Definition 1.10. A cocycle γ on $\overline{\mathfrak{g}}$ is called $\mathcal{V}$-*invariant* if

$$\gamma\big(x(eg), yh\big) + \gamma\big(xg, y(eh)\big) = 0 \qquad (1.6.4)$$

for all $x, y \in \mathfrak{g}$, $e \in \mathcal{V}$, $g, h \in \mathcal{A}$.

The above introduced cocycles are obviously $\mathcal{V}$-invariant.

The following two theorems sum up the results on central extensions of multipoint current Krichever–Novikov algebras. The first of them treats the case when $\mathfrak{g}$ is simple.

Theorem 1.11 ([49, Theorem 3.13, Corollary 3.14]).

(a) *Let $\mathfrak{g}$ be a finite-dimensional simple Lie algebra, then every local cocycle on the current algebra $\overline{\mathfrak{g}} = \mathfrak{g} \otimes \mathcal{A}$ is cohomological to a cocycle given by*

$$\gamma(x \otimes f, y \otimes g) = r \cdot \frac{\beta(x,y)}{2\pi\,\mathrm{i}} \int_{C_S} f dg \qquad (1.6.5)$$

where $r \in \mathbb{C}$, β is a Cartan–Killig form of the Lie algebra $\mathfrak{g}$.

(b) *Every local and $\mathcal{V}$-invariant cocycle coincides with a cocycle of the form* (1.6.5) *for a suitable $r \in \mathbb{C}$ (multipication by r is called rescaling).*

(c) *If $\mathfrak{g}$ is simple then there exists only one, up to equivalence and rescaling, nontrivial almost-graded central extension $\widehat{\mathfrak{g}}$ corresponding to a multipoint higher-genus current algebra $\overline{\mathfrak{g}}$. It is given by a cocycle* (1.6.5).

The second theorem treats the case of a reductive $\mathfrak{g}$. Consider an arbitrary complex reductive Lie algebra $\mathfrak{g}$ and its decomposition

$$\mathfrak{g} = \mathfrak{g}_0 \oplus \mathfrak{g}_1 \oplus \cdots \oplus \mathfrak{g}_M \qquad (1.6.6)$$

into the direct sum of a commutative ideal $\mathfrak{g}_0$ and simple ideals $\mathfrak{g}_1, \ldots, \mathfrak{g}_M$. For the corresponding current algebras we have $\overline{\mathfrak{g}} = \overline{\mathfrak{g}}_0 \oplus \overline{\mathfrak{g}}_1 \oplus \cdots \oplus \overline{\mathfrak{g}}_M$.

Theorem 1.12 ([49, Theorem 3.20]).

(a) *Let $\mathfrak{g}$ be a finite-dimensional reductive Lie algebra over $\mathbb{C}$. For a given local cocycle γ on $\overline{\mathfrak{g}}$ whose restriction onto $\overline{\mathfrak{g}}_0$ is $\mathcal{V}$-invariant there exists a symmetric invariant bilinear form α on $\mathfrak{g}$ such that γ is cohomological to the cocycle*

$$\gamma'_{\alpha,S}(x \otimes f, y \otimes g) = \frac{\alpha(x,y)}{2\pi \mathrm{i}} \int_{C_S} f dg. \tag{1.6.7}$$

Conversely, every such form α defines a local cocycle.

(b) *If a cocycle γ is local and $\mathcal{V}$-invariant on the whole algebra $\overline{\mathfrak{g}}$ then it has the form* (1.6.7).

(c)
$$\dim \mathrm{H}^2_{loc,\mathcal{V}}(\overline{\mathfrak{g}}, \mathbb{C}) = \frac{l(l+1)}{2} + M \tag{1.6.8}$$

where $\mathrm{H}^2_{loc,\mathcal{V}}(\overline{\mathfrak{g}}, \mathbb{C})$ denotes the space of cohomological classes represented by local $\mathcal{V}$-invariant cocycles, $l = \dim \mathfrak{g}_0$.

The statements (a) and (b) of the the theorem reduce the problem of classification of almost-graded central extensions to the classification of invariant symmetric bilinear forms. The statement (c) is obtained by counting the dimension of the latter.

1.7 Central extensions of the Lie algebra $\mathcal{D}^1_{\mathfrak{g}}$

Recall that $\mathcal{D}^1_{\mathfrak{g}} = \overline{\mathfrak{g}} \oplus \mathcal{V}$ with an additional relation $[e, xA] = x(e.A)$; see (1.1.4). There is a short exact sequence of Lie algebras

$$0 \longrightarrow \overline{\mathfrak{g}} \xrightarrow{\ i_1\ } \mathcal{D}^1_{\mathfrak{g}} \xrightarrow{\ p_2\ } \mathcal{V} \longrightarrow 0. \tag{1.7.1}$$

First, we again stress that by almost gradeness of the Lie algebras $\mathcal{V}$ and $\overline{\mathfrak{g}}$, and of the $\mathcal{V}$-module $\mathcal{A}$, the Lie algebra $\mathcal{D}^1_{\mathfrak{g}}$ is also almost-graded. For this algebra, as well as for the above considered algebras, local cocycles and almost-graded central extensions are investigated in [49] (absence of mixing cocycles is observed in [63,64]). By restricting a local cocycle of the algebra $\mathcal{D}^1_{\mathfrak{g}}$ onto the subalgebra $\overline{\mathfrak{g}}$, we obtain a local cocycle on the latter. In the semisimple case we obtain

Theorem 1.13.

(a) *Let $\mathfrak{g}$ be a semisimple Lie algebra, and γ be a local cocycle on $\mathcal{D}^1_{\mathfrak{g}}$. Then there exists a symmetric invariant bilinear form α on $\mathfrak{g}$ such that γ is cohomological to a linear combination of a local cocycle $\gamma_{\alpha,S}$ given by the relation* (1.6.7), *and a local cocycle $\gamma^{(v)}_{S,R}$ on the vector field algebra $\mathcal{V}$ given by* (1.5.8) *(where $C = C_S$).*

(b) *If* $\mathfrak{g}$ *is a simple Lie algebra then the cocycle* $\gamma_{\alpha,S}$ *is a multiple of the standard cocycle* (1.6.5) *on* $\overline{\mathfrak{g}}$.

(c) $\dim \mathrm{H}^2_{loc}(\mathcal{D}^1_{\mathfrak{g}}, \mathbb{C}) = M + 1$ *where* M *is a number of simple ideals in* $\mathfrak{g}$.

It turns out in the reductive case that the restriction of a cocycle on $\mathcal{D}^1_{\mathfrak{g}}$ onto the commutative ideal $\overline{\mathfrak{g}}_0$ is $\mathcal{V}$-invariant. As a generalization of mixing cocycles on $\mathcal{D}^1$, for any linear form $\phi \in \mathfrak{g}^*$ vanishing on $\mathfrak{g}' = [\mathfrak{g}, \mathfrak{g}] = \mathfrak{g}_1 \oplus \cdots \oplus \mathfrak{g}_M$ we obtain a local cocycle

$$\gamma_{\phi,S}(e, xg) := \frac{\phi(x)}{2\pi \mathrm{i}} \int_{C_S} \left(\tilde{e} \cdot g'' + T \cdot (\tilde{e} \cdot g')\right) dz \tag{1.7.2}$$

where T is a meromorphic affine connection having poles only at the points in A, as above.

Theorem 1.14 ([49, Theorem 4.11]).

(a) *Let* $\mathfrak{g}$ *be a finite-dimensional reductive Lie algebra. For any local cocycle* γ *on* $\mathcal{D}^1_{\mathfrak{g}}$, *there is a symmetric invariant bilinear form* α *on* $\mathfrak{g}$, *a linear form* ϕ *on* $\mathfrak{g}$ *vanishing on* $\mathfrak{g}'$, *and* $r \in \mathbb{C}$ *such that* γ *is cohomological to*

$$\gamma' = \gamma_{\alpha,S} + \gamma_{\phi,S} + r\gamma^{(v)}_{S,R}. \tag{1.7.3}$$

The cocycle $\gamma_{\alpha,S}$ *on the current algebra is given by* (1.6.7), *the mixing cocycle* $\gamma_{\phi,S}$ *by* (1.7.2), *and the cocycle* $\gamma^{(v)}_{S,R}$ *on the vector field algebra by* (1.5.8). *Conversely, any such* α, ϕ, $r \in \mathbb{C}$ *define a local cocycle.*

(b) $$\dim \mathrm{H}^2_{loc}\left(\mathcal{D}^1_{\mathfrak{g}}, \mathbb{C}\right) = \frac{l(l+1)}{2} + l + M + 1$$

where $\mathrm{H}^2_{loc}(\mathcal{D}^1_{\mathfrak{g}}, \mathbb{C})$ *is the space of cohomology classes represented by local cocycles,* $l = \dim \mathfrak{g}_0$.

1.8 Local cocycles for $\mathfrak{sl}(n)$ and $\mathfrak{gl}(n)$

Let $\mathfrak{sl}(n)$ be a Lie algebra of traceless matrices $n \times n$ over $\mathbb{C}$. There is the only, up to a scalar factor, invariant symmetric bilinear form on it, namely the Cartan–Killing form $\beta(x, y) = \mathrm{tr}(xy)$. The following statement is implied by Theorems 1.11 and 1.13.

Proposition 1.15.

(a) *Every local cocycle on the current algebra* $\overline{\mathfrak{sl}}(n)$ *is cohomological to the cocycle*

$$\gamma_S(xg, yh) = r \cdot \frac{\mathrm{tr}(xy)}{2\pi \mathrm{i}} \int_{C_S} g\,dh, \quad r \in \mathbb{C}. \tag{1.8.1}$$

(b) *Every $\mathcal{V}$-invariant local cocycle coincides with* (1.8.1) *for a suitable r.*

(c) *Every local cocycle on the differential operator algebra $\mathcal{D}^1_{\mathfrak{sl}(n)}$ is cohomological to a linear combination of* (1.8.1) *and a standard local cocycle $\gamma^{(v)}_{S,R}$ on the vector field algebra. In particular, there is no mixing cocycle.*

Further, let us consider the Lie algebra $\mathfrak{gl}(n)$ of all $n\times n$-matrices over $\mathbb{C}$. It decomposes into a direct sum of a 1-dimensional center $\mathfrak{s}(n)$, and a simple ideal $\mathfrak{sl}(n)$: $\mathfrak{gl}(n) = \mathfrak{s}(n) \oplus \mathfrak{sl}(n) \cong \mathbb{C} \oplus \mathfrak{sl}(n)$ where $\mathfrak{s}(n)$ consists of the scalar $n \times n$-matrices.

The space of symmetric invariant bilinear forms on $\mathfrak{gl}(n)$ is 2-dimensional. A basis is given by

$$\alpha_1(x,y) = \operatorname{tr}(xy), \quad \text{and} \quad \alpha_2(x,y) = \operatorname{tr}(x)\operatorname{tr}(y). \tag{1.8.2}$$

The α_1 is a continuation of the Cartan–Killing form of $\mathfrak{sl}(n)$ onto $\mathfrak{gl}(n)$. It is invariant with respect to $\mathfrak{gl}(n)$. The following statement is implied by Theorems 1.12 and 1.14.

Proposition 1.16.

(a) *A cocycle γ on $\overline{\mathfrak{gl}}(n)$ is local, having a $\mathcal{V}$-invariant restriction onto $\overline{\mathfrak{s}(n)}$ if and only if it is cohomological to a linear combination of the two following cocycles:*

$$\gamma_1(xg, yh) = \frac{\operatorname{tr}(xy)}{2\pi \mathrm{i}} \int_{C_S} g dh, \quad \gamma_2(xg, yh) = \frac{\operatorname{tr}(x)\operatorname{tr}(y)}{2\pi \mathrm{i}} \int_{C_S} g dh. \tag{1.8.3}$$

(b) *If a local cocycle γ on $\overline{\mathfrak{gl}}(n)$ is $\mathcal{V}$-invariant, then it is equal to a linear combination of the cocycles* (1.8.3).

Proposition 1.17.

(a) *Every local cocycle γ on $\mathcal{D}^1_{\mathfrak{gl}(n)}$ is cohomological to a linear combination of the cocycles γ_1 and γ_2 defined by the relations* (1.8.3), *the mixing cocycle*

$$\gamma_{3,T}(e, xg) = \frac{\operatorname{tr}(x)}{2\pi \mathrm{i}} \int_{C_S} (\tilde{e}g'' + T\tilde{e}g')dz, \tag{1.8.4}$$

and the standard local cocycle $\gamma^{(v)}_{S,R}$ of the vector field algebra, i.e.,

$$\gamma = r_1\gamma_1 + r_2\gamma_2 + r_3\gamma_{3,T} + r_4\gamma^{(v)}_{S,R} + \text{coboundary} \tag{1.8.5}$$

for suitable $r_1, r_2, r_3, r_4 \in \mathbb{C}$.

(b) *If the cocycle γ is local, its restriction to $\overline{\mathfrak{gl}}(n)$ is $\mathcal{V}$-invariant, and $r_3, r_4 \neq 0$, then there exists an affine connection T and a projective connection R holomorphic outside A such that the cocycle γ has the form* (1.8.5) *with a zero coboundary.*

(c) $\dim \mathrm{H}^2_{loc}(\mathcal{D}^1_{\mathfrak{gl}(n)}, \mathbb{C}) = 4$.

It turns out that the restrictions of the cocycles of the differential operator algebra to $\overline{\mathfrak{gl}}(n)$ are cohomological to $\mathcal{V}$-invariant cocycles. Moreover, the restrictions to $\mathfrak{s}(n)$ *are* $\mathcal{V}$-invariant [49, Prop. 4.10].

Chapter 2

Fermion representations and Sugawara construction

A complete representation theory of Krichever–Novikov algebras has not yet been created, but a certain minimum enabling us to give some applications is known. There is a certain understanding, inspired by the orbit method, of what a generic representation of an affine Krichever–Novikov algebra depends on. In this chapter we present the construction of fermion representations which realizes this understanding. We do it for both affine and Virasoro-type Krichever–Novikov algebras.

We show that one of the most powerful representation theory tools, especially important for applications in Conformal Field Theory, geometry, combinatorics, and the theory of automorphic forms, namely the Sugawara construction, is available in the representation theory of Krichever–Novikov algebras. The content of the present chapter is quite often used in the applications described in Chapters 3 and 6.

The construction of fermion representations (semiinfinite monomials in another terminology) is introduced in [6] for Kac–Moody and Virasoro algebras, and in [26–28] for Krichever–Novikov algebras. In [62] we introduced the holomorphic vector bundles into the construction. See also [26,50,52,53,58,60,61,66–68] for other approaches to the representations of Krichever–Novikov algebras, and for further details.

2.1 Admissible representations and holomorphic bundles

Let V be a vector space, $r : \overline{\mathfrak{g}} \to End(V)$ be a linear mapping such that $r([X,Y]) = [r(X), r(Y)] + \gamma(X,Y)Id$ where γ is a 2-cocycle on $\overline{\mathfrak{g}}$, $X, Y \in \overline{\mathfrak{g}}$ are arbitrary, $Id \in End(v)$ is the unit. Then r is called a *projective representation* of $\overline{\mathfrak{g}}$ in V. V is called the space of this representation, the pair (V, r) is called a *projective $\overline{\mathfrak{g}}$-module*.

The term projective representations of a Lie algebra is another way of calling representations of its central extensions, not specifying the cocycles of the extensions.

If the $\overline{\mathfrak{g}}$-action on V given by r satisfies the relations (1.4.2) then the projective representation r is called almost-graded.

A $v_0 \in V$ is called a *singular vector (vacuum vector, vacuum)* if $Xv = 0$ for all X such that $\deg X > 0$ or $X \in \mathfrak{N}_+$ where $\mathfrak{N}_+ \subset \mathfrak{g}$ is the upper nilpotent subalgebra. We tell that v_0 generates V if $V = U(\overline{\mathfrak{g}})v_0$ where $U(\overline{\mathfrak{g}})$ is the universal enveloping algebra of $\overline{\mathfrak{g}}$.

The same terminology is applicable to Virasoro-type algebras. In this case vacuum is defined as an element annihilated by $\mathcal{V}_+$.

For physical applications we need *vacuum modules*. By this we mean a projective almost-graded module generated by vacuum. However, some important constructions below, in particular the Sugawara construction, are valid in a more general mathematical context of admissible representations. An almost-graded (projective) module is called *admissible* if every element of the representation space is annihilated by all elements of sufficiently large degree in the Lie algebra.

Example 2.1. Any fermion representation constructed below is admissible. The adjoint representation of $\overline{\mathfrak{g}}$ is not admissible.

There is a conventional procedure [6, 16] making it possible, given an arbitrary graded module, to construct the corresponding admissible representation called its semiinfinite degree, or wedge representation. In this way fermion representations were constructed in [26, 27] for $\mathcal{V}$ and $\mathcal{A}$. The authors first constructed certain natural modules over those algebras, which were named Klebsh–Gordon–Baker–Akhieser modules by them, and passed to their semiinfinite degree. For that purpose the property of almost-gradeness was successfully used. The multipoint version of the construction is developed in [44].

For the affine Krichever–Novikov algebras the fermion representations are introduced in [62]. They are constructed following the same scheme, except that we used the space of the meromorphic sections of a holomorphic vector bundle on Σ as a starting point. We denote this space as $\mathcal{F}$.

It is by no means accidental that holomorphic vector bundles occur in the representation theory of affine Krichever–Novikov algebras. The first reason is given by the Kirillov orbital principle. Essentially, this states that generic representations of a Lie algebra are in one-to-one correspondence with the coadjoint orbits of the algebra. In [59, 67, 68], it is shown that such an orbit, in turn, is given by an equivalence class of representations of the fundamental group of the punctured Riemann surface. The relation of this data to holomorphic vector bundles on Riemann surfaces is well-known due to the work of Narasimhan and Seshadry on the classification of the moduli of stable holomorphic vector bundles on Riemann surfaces, of Deligne, in the context of the relation between holomorphic bundles and local systems, and of other authors.

The second reason for the occurrence of the holomorphic vector bundles in the representation theory of Krichever–Novikov algebras is given by the theory of commutative rings of difference operators. As shown in [22, 30, 31], the classification of such commutative rings is closely related to the holomorphic vector bundles on algebraic curves. On the other hand, the almost-graded structure on the modules such that the dimension of a homogeneous component is bounded enables one to regard the representation operators as difference operators. Below we heavily use the bases in $\mathcal{F}$ constructed by I. M. Krichever and S. P. Novikov [31] using earlier results [23, 25] due to the same authors, and to A. N. Tyurin [75]. In Section 2.3 we present this construction, extending the presentation of [31] somewhat. The idea of the relationship between Krichever–Novikov algebras and the soliton theory was one of the starting points of [26–28]. It is confirmed again in the present book by the appearance of Tyurin parameters both in the construction of fermion representations and in the theory of Lax operators with the spectral parameter on a Riemann surface, and also by the appearance of the algebra $\mathcal{A}$ in the Conformal Field Theory related to an integrable system in Chapter 6.

2.2 Holomorphic bundles in the Tyurin parametrization

As already mentioned, fermion representations are related to holomorphic vector bundles on the Riemann surface Σ. The aim of this section is to give an effective description of the moduli space of such bundles.

Let us consider a holomorphic rank r degree gr bundle F on Σ. By the Riemann-Roch theorem, in a generic position, F has r holomorphic sections $\Psi_1, \ldots, \Psi_r$ constituting a base in the fibre over each point, except over gr mutually different points. We denote these degeneration points $\gamma_1, \ldots, \gamma_{gr}$, and refer to $D = \sum_{i=1}^{gr} \gamma_i$ as to the Tyurin divisor of the bundle F. Following [23, 25], the set $\Psi_1, \ldots, \Psi_r$ is called *framing*, and the bundle with a given framing is called a *framed bundle*.

Given a local trivialization of F, it is possible to consider the sections $\Psi_1, \ldots, \Psi_r$ as local vector-valued functions. Let us think of them as columns of a matrix Ψ. Thus Ψ is invertible everywhere except at the points γ_i, $i = 1, \ldots, gr$ where $\det \Psi$ has simple poles: $\det \Psi(\gamma_i) = 0$, $(\det \Psi)'(\gamma_i) \neq 0$. Let us assume that $\operatorname{rank} \Psi(\gamma_i) = r - 1$, $i = 1, \ldots, gr$. We may do this in a generic position. Then for any $i = 1, \ldots, gr$ there exists a unique, up to a proportionality, nontrivial solution to the system of linear equations $\Psi(\gamma_i)\alpha_i = 0$ where $\alpha_i = (\alpha_{i1}, \ldots, \alpha_{ir})^t$, the upper t means transposition, i.e., α_i is regarded as a vector-column. The Tyurin divisor and the vectors α_i, $i = 1, \ldots, gr$ are called the Tyurin parameters of the bundle F [23, 25]. The framing is defined up to an action of the group $GL(r)$ on Ψ by the *right* multiplication, while the gluing functions of the bundle act by the *left* multiplication. We see that the first $GL(r)$-action commutes with the action of gluing functions; thus it sends sections to

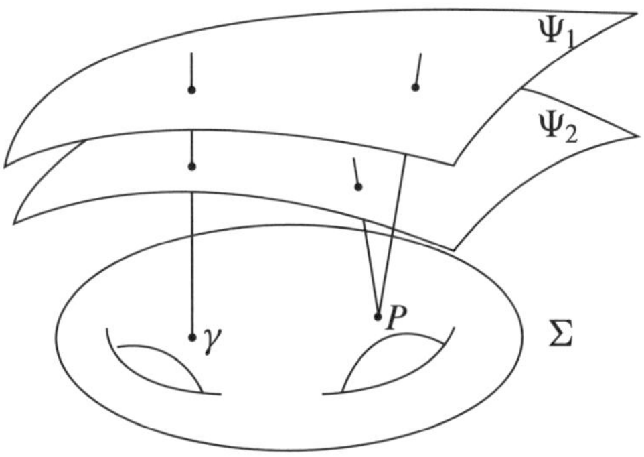

Figure 2.1. Tyurin parameters for a rank 2 vector bundle. Here γ is a Tyurin point, P is a generic point. In the fiber over γ, the intersection points are linear dependent, and over P linear independent.

sections. Hence the Tyurin parameters are defined up to proportionality and the *left* action of $GL(r)$. Observe that equivalent framed bundles have *the same* set $\gamma_1, \dots, \gamma_{gr}$ while for the bundles considered without any framing only the *equivalence class* of the divisor appears as an invariant. According to [75] the Tyurin parameters uniquely determine the holomorphic vector bundle up to equivalence.

There is a way, proposed in [23,25], to represent the space of the meromorphic sections of the holomorphic bundle F as a certain function space on the Riemann surface. Let S be such a meromorphic section holomorphic outside the points $P_1, \dots, P_N, P_\infty$. Consider a mapping ψ taking S to the solution of the equation $S = \Psi f$:

$$\psi : S \to f.$$

Proposition 2.2 ([23,25]).

(1) *The mapping ψ is an isomorphism between the space of meromorphic sections of the bundle F and the space of meromorphic vector-valued functions $f = (f_1, \dots, f_r)^t$ on Σ having at most simple poles at the points $\gamma_1, \dots, \gamma_{gr} \in |D|$, and satisfying the condition*

$$f(z) = \frac{\lambda\alpha}{z - z_\gamma} + O(1) \tag{2.2.1}$$

for every $\gamma \in |D|$ where z is a local coordinate in the neighborhood of γ, z_γ is the coordinate of the γ itself, $\lambda \in \mathbb{C}$, $|D|$ is the support of the divisor D.

(2) *If $S = \Psi f$ then* $\mathrm{ord}_P\, S = \mathrm{ord}_P f$ *for every $P \in \Sigma$, $P \notin |D|$.*

Proof. (1) Outside $|D|$, the matrix Ψ is nondegenerate; hence given S we can define f by the relation $f = \Psi^{-1} S$. By Cramer's rule

$$f_j = \frac{\det(\Psi_1, \dots, \Psi_{j-1}, S, \Psi_{j+1}, \dots, \Psi_r)}{\det \Psi},$$

i.e., is meromorphic, having poles only at the points in $|D|$. Since $\Psi_1, \dots, \Psi_r$ are holomorphic at the points in D, and $\det \Psi$ has at most simple zeroes there, f has at most simple poles there.

In the local coordinates in the neighborhood of a $\gamma_i \in |D|$, the relation $S = \Psi f$ reads as $S(z) = \Psi(\gamma)(\mathrm{res}_{\gamma_i} f) z^{-1} + O(1)$ where we take $z_{\gamma_i} = 0$ for brevity. The left-hand side of that relation is holomorphic, hence the residue of the right-hand side is equal to zero: $\Psi(\gamma_i)(\mathrm{res}_{\gamma_i} f) = 0$. This is just the system defining α_i. By assumption, $\mathrm{rank}\, \Psi(\gamma_i) = r - 1$. Hence $\mathrm{res}_{\gamma_i} f$ and α_i are proportional.

(2) is immediately implied by invertibility of Ψ outside $|D|$. □

Denote the ψ-image of the space of meromorphic sections holomorphic outside $P_1, \dots, P_N, P_\infty$ by $\mathcal{F}$. By Proposition 2.2(2) $\mathcal{F}$ consists of the meromorphic vector-valued functions holomorphic outside $P_1, \dots, P_N, P_\infty$ and $|D|$, and satisfying the condition (2.2.1) at the points in $|D|$.

$\mathcal{F}$ is a natural $\mathcal{A}$-module where functions act on the elements of $\mathcal{F}$ by multiplication.

2.3 Krichever–Novikov bases for holomorphic vector bundles

In this section we introduce a certain base in $\mathcal{F}$. It gives an almost-graded structure on $\mathcal{F}$ and makes it possible to define the space of semiinfinite forms on $\mathcal{F}$.

Let us start with case of two marked points $P_\pm$. For an arbitrary pare of integers n, j $(0 \le j < r)$, let us construct a vector-valued function $\psi_{n,j} \in \mathcal{F}$. The n is called *degree* of $\psi_{n,j}$. The function $\psi_{n,j}$ is defined by its asymptotic behavior at the points $P_\pm$. Let us think of a $\psi_{n,j}$ as of vector-column, form the matrix Ψ_n of those columns, and require that

$$\Psi_n(z_+) = z_+^n \sum_{s=0}^{\infty} \xi_{n,s}^+ z_+^s, \qquad \xi_{n,0}^+ = \begin{pmatrix} 1 & * & \dots & * \\ 0 & 1 & \dots & * \\ & \dots\dots & & \\ 0 & 0 & \dots & 1 \end{pmatrix}, \tag{2.3.1}$$

and

$$\Psi_n(z_-) = z_-^{-n} \sum_{s=0}^{\infty} \xi_{n,s}^- z_-^s, \qquad \xi_{n,0}^- = \begin{pmatrix} * & 0 & \dots & 0 \\ * & * & \dots & 0 \\ & \dots\dots & & \\ * & * & \dots & * \end{pmatrix} \tag{2.3.2}$$

where $z_\pm$ are local coordinates at the points $P_\pm$, respectively, and stars denote arbitrary complex numbers. The matrix Ψ_n has has a zero of order n at one of the points $P_\pm$, and a pole of the same order at the other one. Its determinant has gr simple poles at the points of the divisor D, and some zero divisor outside $|D| \cup \{P_\pm\}$.

Let us call $\{\psi_{n,j}\}$ the *Krichever–Novikov base* in $\mathcal{F}$.

The case $r = 1$ is exceptional and requires a special approach. In the most important cases of the rank 1 trivial bundle, and the tangent bundle, this was already done in Section 1.3.

Proposition 2.3 ([62]).

(1) *There exists a unique matrix-valued function Ψ_n satisfying the conditions* (2.3.1), (2.3.2).

(2) *The dimension of the space generated by the vector-valued functions $\psi_{n,j}$ for a given n is equal to r.*

Proof. It is clear that assertions 1 and 2 are equivalent. In the part related to the matrices $\xi_{n,0}^\pm$, conditions (2.3.1) and (2.3.2) are the normalization conditions that uniquely select the vector functions $\psi_{n,j}$ in the r-dimensional space.

By the Riemann–Roch theorem, the dimension of the space of functions with gr poles at the points of the divisor D with the orders $\pm n$ at the points $P_{\pm}$, respectively, is equal to $gr - g + 1$. Since our vector functions are of dimension r, we find that the dimension of the corresponding space is $r(gr - g + 1)$. However, these functions satisfy the $(r - 1)gr$ relations (2.2.1). Indeed, there are r constrains on the residue containing 1 free parameter λ for every $\gamma \in |D|$. Therefore, the dimension of the function space under consideration is equal to $r(gr - g + 1) - (r - 1)gr = r$. □

The algebra $\mathcal{A} = \mathcal{A}(\Sigma, P_{\pm})$ has a natural action in the space $\mathcal{F}$ given by multiplication of the elements of the latter by functions.

Proposition 2.4 ([62]). *The $\mathcal{A}$-action on $\mathcal{F}$ is almost-graded. For the basis elements $A_m \in \mathcal{A}$ and $\psi_{n,j} \in \mathcal{F}$ we have*

$$A_m \psi_{n,j} = \sum_{k=m+n}^{m+n+\bar{g}} \sum_{j'=0}^{l-1} C_{m,n,j}^{k,j'} \psi_{k,j'} \tag{2.3.3}$$

where $\bar{g} = g + 1$ for $-g \le m \le 0$ and $\bar{g} = g$ otherwise, $C_{m,n,j}^{k,j'} \in \mathbb{C}$.

Proof. The A_m is given by the asymptotic behavior (1.3.6). The product $A_m \Psi_n$ has the order z_+^{n+m} at the point P_+ and the order $z_-^{-n-m-\bar{g}}$ at the point P_-. It is sufficient for the validity of relations (2.3.3) for some values of the constants $C_{m,n,j}^{k,j'}$. In particular, the diversity of the grading with respect to the index n is equal to g except in the case $-g \le m \le 0$ when it is equal to $g + 1$, and hence it is bounded. This proves that the action is almost-graded. □

In contrast with the algebra $\mathcal{A}$, the Lie algebra $\mathfrak{g}$ admits no natural action on sections of the bundle since it would destroy the relation (2.2.1) (we will revisit this question in Chapter 4). However, suppose that a representation τ of the Lie algebra $\mathfrak{g}$ in an l-dimensional space V_τ is given. Assume that we have fixed a base in V_τ, or, equivalently, an isomorphism $V_\tau \cong \mathbb{C}^l$. Let $\tau(x) = (x_{i'}^i)$ where $x \in \mathfrak{g}$, $(x_{i'}^i) \in \mathfrak{gl}(l)$ is the matrix representing x, and the indices i, i' range $\{1, \dots, l\}$. Let us associate every basis element $\psi_{n,j}$ $(n \in \mathbb{Z},\ j = 0, \dots, r - 1)$ with the corresponding set of basis elements $\{\psi_{n,j}^i : i = 1, \dots, l\}$ of the space $\mathcal{F}_\tau = \mathcal{F} \otimes V_\tau$. Define an action of the current algebra $\mathfrak{g} \otimes \mathcal{A}$ in $\mathcal{F}_\tau$ as follows:

$$(xA_m)\psi_{n,j}^i = \sum_{k=m+n}^{m+n+\bar{g}} \sum_{j'=0}^{r-1} \sum_{i'=1}^{l} C_{m,n,j}^{k,j'} x_{i'}^i \psi_{k,j'}^{i'}. \tag{2.3.4}$$

This is nothing but a usual action of the tensor product which looks as follows in the more invariant form:

$$(xA)(\psi \otimes v) = A\psi \otimes xv, \qquad A \in \mathcal{A}, \quad \psi \in \mathcal{F}, \quad v \in V_\tau. \tag{2.3.5}$$

It follows from Proposition 2.3 that the action (2.3.4) is almost-graded, and its diversity of grading with respect to the index n is equal to g. The corresponding action of $\mathfrak{g}$ does not change the indices n, j; for given values of n, j, the role of the highest weight vector is played by $\psi^l_{n,j}$.

Now, combining the approaches of this section and Section 1.3, we shall introduce Krichever–Novikov bases for an arbitrary N where $N+1$ is a total number of marked points.

From now on, let l stand for the rank of $\mathfrak{g}$.

For any triple of integers $n \in \mathbb{Z}$, $j = 0, 1, \ldots r-1$ and $p = 1, \ldots, N$, let $\psi_{n,j,p} = (\psi^i_{n,j,p})$ be a vector function ($i = 0, 1, \ldots, r-1$) in $\mathcal{F}$. This function is specified by its asymptotic behavior at the points of A which is assumed to be as follows:

$$\psi^i_{n,j,p}(z_q) = z_q^{n+1-\delta_{p,q}}\left(\xi^i_{npqj} + O(z_q)\right) \tag{2.3.6}$$

where z_q is a local coordinate at P_q, $q \in \{1, \ldots, N\}$, $\delta_{p,q}$ is the Kronecker symbol, ξ_{npqj} are the complex numbers such that $\xi^i_{nppi} = 1$, $\xi^i_{nppj} = 0, i > j$;

$$\psi^i_{n,j,p}(z_\infty) = z_\infty^{-nN-N+1}\left(\xi^i_{np\infty j} + O(z_\infty)\right) \tag{2.3.7}$$

where $\xi^i_{np\infty j} = 0, i < j$.

For a given n, the space spanned by vector-valued functions $\psi_{n,j,p}$ is referred to as the space of degree n functions.

Proposition 2.5 ([67]).

(1) *There exists a unique vector-valued function $\psi_{n,j,p}$ satisfying conditions* (2.3.6) *and* (2.3.7).

(2) *For a given n, the dimension of the space spanned by the vector-valued functions $\psi_{n,j,p}$ is equal to rN.*

Proof. The proof is quite similar to the proof of Proposition 2.3. Introduce the divisor $D_{n,p} = D + nP_p + \Sigma_{q\neq p}(n+1)P_q + (-nN - N + 1)P_\infty$, hence deg $D_{n,p} =$ deg $D = -rg$. Let $(D_{n,p})$ denote the space of (scalar-valued) meromorphic functions with a divisor not less than $D_{n,p}$, then $\dim(D_{n,p}) = rg - g + 1$. For the analogous space of functions taking values in an r-dimensional vector space (denote it by $(D_{n,p})_r$), we have $\dim (D_{n,p})_r = r(rg - g + 1)$. Such functions in $\mathcal{F}$ satisfy the $(r-1)gr$ Tjurin relations. Therefore, $\dim ((D_{n,p})_r \cap \mathcal{F}) = r(gr - g + 1) - (r-1)gr = r$.

Observe that $\psi_{n,j,p} \in (D_{n,p})_r \cap \mathcal{F}$ for any j. For a given j, this function is distinguished in $(D_{n,p})_r$ by exactly r (normalizing) conditions on matrices ξ. Thus, $\psi_{n,j,p}$ is uniquely defined and the assertion 1 is proven.

Further on, observe that, for a given n, the space of all degree n functions in $\mathcal{F}$ is just a direct sum of its subspaces $(D_{n,p})_r \cap \mathcal{F}$, $p = 1, \ldots, N$. Hence, its dimension is equal to rN. □

2.4 Fermion representations of affine algebras

We will proceed by considering the two-point case, for simplicity. Let us number the elements $\psi^i_{n,j}$ in the lexicographically ascending order of the triples (n, j, i). Let $N = N(n, j, i)$ be the index of a triple (n, j, i) (the reader should not confuse it with the number of marked points; we will not use N for that purpose in this section). Here, n is an arbitrary integer, $j = 0, \ldots, r-1$, $i = 0, \ldots, l-1$, hence

$$N = nrl + jl + i. \tag{2.4.1}$$

Set $\psi_N = \psi^i_{n,j}$. In the case of several marked points, when we have basis vectors $\psi^i_{n,j,p}$, we think of ψ_N as having an additional index p labelling the marked points. Then we lexicographically order the pairs (N, p) considering N to have priority. Under this convention, the additional index does not affect the presentation in the remainder of this section.

Let us introduce the fermion representation $\pi_{F,\tau}$ corresponding to F, τ as follows. The representation space is spanned over $\mathbb{C}$ by formal expressions (referred to as *semi-infinite monomials*) of the form $\psi_{N_0} \wedge \psi_{N_1} \wedge \ldots$ where $N_0 < N_1 < \cdots$. It is required that a semiinfinite monomial changes its sign under the transposition of ψ_N and $\psi_{N'}$, and one has $N_k = k + m$ for sufficiently large k where m is an integer. Following [17], m is called the *charge* of the monomial. Let us denote the space of all semiinfinite monomials by $\mathcal{F}_\tau^{\infty/2}$, and the space of monomials of charge m by $\mathcal{F}_{\tau,m}^{\infty/2}$.

By the *degree of a monomial* ψ of charge m one means

$$\deg \psi = \sum_{k=0}^{\infty} (N_k - k - m). \tag{2.4.2}$$

Example 2.6. The charge of the monomial $\psi_{-5} \wedge \psi_{-3} \wedge \psi_1 \wedge \psi_2 \wedge \psi_3 \wedge \cdots$ is equal to -1, and its degree is equal to -7.

The projective representation $\pi_{F,\tau}$ of the Lie algebra $\overline{\mathfrak{g}}$ in the space $\mathcal{F}_\tau^{\infty/2}$ is defined in two steps. First, observe that by (2.3.4) every element of $\overline{\mathfrak{g}}$ acts in $\mathcal{F}$ by a linear transformation in an almost-graded way. It was mentioned on page 25 that the diversity of grading with respect to the index n is equal to g. By (2.4.1) the diversity of grading with respect to the index N is finite and equal to rlg. The finiteness precisely means that in the basis $\{\psi_N\}$, the action of any element of the Lie algebra $\overline{\mathfrak{g}}$ in $\mathcal{F}$ is given by an infinite matrix with finitely many nonzero diagonals. Following [15], we denote the algebra of these matrices by $\mathfrak{a}_\infty$. Thus, to the basis $\{\psi_N\}$ there corresponds an embedding of $\overline{\mathfrak{g}}$ in the algebra $\mathfrak{a}_\infty$. It suffices to define an action of $\mathfrak{a}_\infty$ in $\mathcal{F}_\tau^{\infty/2}$, and then, by virtue of the embedding, we obtain a representation of $\overline{\mathfrak{g}}$ in the same space.

Remark 2.7. The $\mathfrak{a}_\infty$ is the same as the algebra of difference operators in a 1-dimensional lattice. A matrix with $k+1$ nonzero diagonals defines the operator of order k. This observation establishes a relation to [31].

The algebra $\mathfrak{a}_\infty$ possesses a standard action in $\mathcal{F}_\tau^{\infty/2}$ induced by its action in $\mathcal{F}$ considered above. Recall from [17,62] that the action of the basis elements $E_{IJ} \in \mathfrak{a}_\infty$ on semiinfinite monomials $\psi = \psi_{I_0} \wedge \psi_{I_1} \wedge \cdots$ is given by the Leibniz rule:

$$r(E_{IJ})\psi = (E_{IJ}\psi_{I_0}) \wedge \psi_{I_1} \wedge \cdots + \psi_{I_0} \wedge (E_{IJ}\psi_{I_1}) \wedge \cdots + \cdots . \qquad (2.4.3)$$

Indeed, the action of $\mathfrak{a}_\infty$ conserves the charge of a monomial, since infinite tails of the monomials on both hand sides of the relation (2.4.3) are the same. Hence, the fermion space decomposes to a direct sum of $\overline{\mathfrak{g}}$-invariant subspaces corresponding to different values of the charge. For $\mathfrak{g} = \mathfrak{gl}(l)$, the subspace of charge m is generated by the monomial $|0\rangle = \psi_m \wedge \psi_{m+1} \wedge \cdots$ over the universal enveloping algebra $\mathcal{U}(\widehat{\mathfrak{g}})$. This monomial is referred to as *vacuum monomial*, or just *vacuum*, of charge m. Every vacuum monomial possesses the following property: $\pi_{F,\tau}(xA)\,|0\rangle = 0$ for $A \in \mathcal{A}_+$, and also for $A = 1$ and each strictly upper triangular matrix $x \in \mathfrak{g}$.

By virtue of the condition $I_k = k + m$ (for sufficiently large k) the action (2.4.3) is well defined for $I \neq J$. For $I = J$, there are infinitely many summands on the right hand side of (2.4.3). In this case the standard regularization procedure [17, 62] must be applied: we set $\widehat{r}(E_{IJ}) = r(E_{IJ}) - Id$ for $I = J > 0$, and $\widehat{r}(E_{IJ}) = r(E_{IJ})$ otherwise. The commutation relations between the operators $\widehat{r}(E_{IJ})$ are the same as between the corresponding $r(E_{IJ})$, up to the additional scalar operators occurring on the right-hand side. Hence the regularization results in a projective representation $\pi_{F,\tau}$ of the Lie algebra $\mathfrak{a}_\infty$, and also of the Lie algebra $\mathfrak{g} \otimes \mathcal{A}$. For the latter, this is an almost-graded representation, as it follows from the almost-grading of the action (2.3.4) in $\mathcal{F}$. Below, we need the following lemma.

Lemma 2.8 ([17]). *The cocycle of the fermion representation of the Lie algebra $\mathfrak{a}_\infty$ satisfies the following relations:*

$$\begin{aligned} &\widehat{\gamma}(E_{IJ}, E_{JI}) = -\widehat{\gamma}(E_{JI}, E_{IJ}) = 1, && \text{if } I \le 0, J > 0, \\ &\widehat{\gamma}(E_{IJ}, E_{MN}) = 0 && \text{otherwise.} \end{aligned} \qquad (2.4.4)$$

Consider the question whether $\pi_{F,\tau}$ is a representation of the affine algebra $\widehat{\mathfrak{g}}$. Let $\widehat{\gamma}$ be the cocycle of the representation $\pi_{F,\tau}$. In terms of $\widehat{\gamma}$, the question is whether or not $\widehat{\gamma}$ is of the form (1.6.2), up to proportionality.

Theorem 2.9. *If $\mathfrak{g}$ is a simple Lie algebra, then the cocycle of the fermion representation is proportional to* (1.6.2).

It would be a short task to prove this theorem if we could refer to some uniqueness result such as Theorem 1.11b. Unfortunately, we do not know a priori that the cocycle $\widehat{\gamma}$ is $\mathcal{V}$-invariant. Nevertheless, it is. To prove $\mathcal{V}$-invariance, we construct a natural $\mathcal{V}$-action in the representation space $\mathcal{F}_\tau^{\infty/2}$. We show that together with the $\overline{\mathfrak{g}}$-action it gives a projective representation of $\mathcal{D}^1_{\mathfrak{g}}$. For commutative current algebras it is easy to prove that if the cocycle on currents is a restriction of a cocycle on the first order differential operators, then it is $\mathcal{V}$-invariant. Unfortunately, $\overline{\mathfrak{g}}$ is not commutative. Thus, our plan is as follows: we split $\widehat{\gamma}$ to the product of an invariant quadratic form on $\mathfrak{g}$ and a cocycle on $\mathcal{A}$ (Lemma 2.10), and prove that the last is $\mathcal{V}$-invariant.

In the remainder of this section we prove Lemma 2.10, delaying the proof of Theorem 2.9 until Section 2.6.

Lemma 2.10. *The cocycle of the fermion representation has the form*

$$\widehat{\gamma}(xA, yB) = \alpha(x, y)\,\gamma(A, B) \tag{2.4.5}$$

where γ is a cocycle on the Lie algebra $\mathcal{A}$, α is an invariant symmetric bilinear form on $\mathfrak{g}$. In particular, if $\mathfrak{g} = \mathfrak{gl}(l)$, then $\alpha(x, y) = \operatorname{tr}(xy)$.

Proof. Let $e_{ij} \in \mathfrak{gl}(l)$ be the matrix with only one nonzero entry which is equal to 1 and whose indices are i, j, Let E^I_J be a similar element in $\mathfrak{a}_\infty$ (which was denoted above by E_{IJ}) where I stays for the index of the row, and J for the index of the column. Denote the representation operator of the element $e_{ij}A_m$ by $e_{ij}(A_m)$.

By (2.3.4) we have

$$e_{ij}(A_m)\psi^j_{n,k} = \sum_{n'=m+n}^{m+n+\bar{g}} \sum_{k'=0}^{r-1} C^{n',k'}_{m,n,k} \psi^i_{n',k'}, \tag{2.4.6}$$

and $e_{ij}(A_m)$ annihilates $\psi^{j'}_{n,k}$ if $j \neq j'$. Passing from the three-index labelling of ψ-functions to the one-index (2.4.1) we obtain

$$e_{ij}(A_m)\psi_{nrl+kl+j} = \sum_{n'=m+n}^{m+n+\bar{g}} \sum_{k'=0}^{r-1} C^{n',k'}_{m,n,k} \psi_{n'rl+k'l+i} \tag{2.4.7}$$

from (2.4.6). From the last relation we see that $C^{n',k'}_{m,n,k}$ is just the matrix element of the operator $e_{ij}(A_m)$ corresponding to the row $n'rl + k'l + i$, and column $nrl + kl + j$. Hence this relation is written:

$$e_{ij}(A_m) = \sum_{n\in\mathbb{Z}} \sum_{k=0}^{r-1} \sum_{n'=m+n}^{m+n+\bar{g}} \sum_{k'=0}^{r-1} C^{n',k'}_{m,n,k} E^{n'rl+k'l+i}_{nrl+kl+j}. \tag{2.4.8}$$

For arbitrary six indices $i, \bar{i}, j, \bar{j}, m, \bar{m}$ we have

$$\widehat{\gamma}(e_{ij}A_m, e_{\bar{i}\bar{j}}A_{\bar{m}}) = \sum_{n,\bar{n}\in\mathbb{Z}} \sum_{k,\bar{k}=0}^{r-1} \sum_{n'=m+n}^{m+n+g} \sum_{\bar{n}'=\bar{m}+\bar{n}}^{\bar{m}+\bar{n}+g} \sum_{k',\bar{k}'=0}^{r-1} C^{n',k'}_{m,n,k} C^{\bar{n}',\bar{k}'}_{\bar{m},\bar{n},\bar{k}} \times \widehat{\gamma}\left(E^{n'rl+k'l+i}_{nrl+kl+j}, E^{\bar{n}'rl+\bar{k}'l+\bar{i}}_{\bar{n}rl+\bar{k}l+\bar{j}}\right). \tag{2.4.9}$$

By Lemma 2.8 the value of the cocycle on the right-hand side of (2.4.9) is equal to 1 by coincidence of the upper index of the first operator with the lower index of the second one, and vice verse, and zero otherwise. Hence

$$\widehat{\gamma}\left(E^{n'rl+k'l+i}_{nrl+kl+j}, E^{\bar{n}'rl+\bar{k}'l+\bar{i}}_{\bar{n}rl+\bar{k}l+\bar{j}}\right) = \delta_{i\bar{j}}\delta_{j\bar{i}}\delta_{n'\bar{n}}\delta_{n\bar{n}'}\delta_{k'\bar{k}}\delta_{k\bar{k}'}. \tag{2.4.10}$$

By substituting the just obtained value of the cocycle into (2.4.9) (and denoting the multiple summing there by the single symbol $\sum$) we obtain

$$\widehat{\gamma}(e_{ij}A_m, e_{\bar{i}\bar{j}}A_{\bar{m}}) = \delta_{i\bar{j}}\delta_{j\bar{i}} \sum C^{n',k'}_{m,n,k} C^{\bar{n}',\bar{k}'}_{\bar{m},\bar{n},\bar{k}} \delta_{n'\bar{n}}\delta_{n\bar{n}'}\delta_{k'\bar{k}}\delta_{k\bar{k}'} \tag{2.4.11}$$

(summing over all pairs of repeating indices).

Observe that the sum in (2.4.11) depends only on m, $\bar{m}$. Denote this sum by $\gamma(A_m, A_{\bar{m}})$. The factor before the sum is equal to $\operatorname{tr}(e_{ij}e_{\bar{i}\bar{j}})$. Hence

$$\widehat{\gamma}(e_{ij}A_m, e_{\bar{i}\bar{j}}A_{\bar{m}}) = \operatorname{tr}(e_{ij}e_{\bar{i}\bar{j}})\gamma(A_m, A_{\bar{m}}).$$

By bilinearity, for arbitrary $x, y \in \mathfrak{gl}(l)$ we finally obtain

$$\widehat{\gamma}(xA_m, yA_{\bar{m}}) = \operatorname{tr}(xy)\gamma(A_m, A_{\bar{m}}), \tag{2.4.12}$$

which proves the lemma for $\mathfrak{g} = \mathfrak{gl}(l)$.

If $\mathfrak{g}$ is an arbitrary reductive Lie algebra, and τ its representation in $\mathfrak{gl}(l)$, then the fermion representation is obtained as a composition of τ and the fermion representation of $\mathfrak{gl}(l)$. For this reason the cocycle is equal to

$$\widehat{\gamma}(\tau(x)A_m, \tau(y)A_{\bar{m}}) = \operatorname{tr}(\tau(x)\tau(y))\gamma(A_m, A_{\bar{m}}),$$

which coincides with (2.4.5) for $\alpha(x, y) = \operatorname{tr}(\tau(x)\tau(y))$. □

2.5 Verma modules for affine algebras

In this section we construct one more class of representations of the affine Krichever–Novikov algebras introduced in [52]. In the rational case, and for two marked points, they reduce to conventional Verma modules. For this reason, they are also referred to as Verma modules. They depend on a smaller number of parameters than fermion representations. Nevertheless they are also used in conformal field theory [52].

Let $\mathfrak{g}$ be a simple finite-dimensional Lie algebra, $\overline{\mathfrak{g}}_0 \subset \widehat{\mathfrak{g}}$ be a linear space of the elements of degree 0, i.e., $\overline{\mathfrak{g}}_0 = \bigoplus_{p=1}^{N} \mathfrak{g} \otimes A_{0,p}$. Let $\widehat{\mathfrak{g}}_+ = \overline{\mathfrak{g}}_+ \subset \overline{\mathfrak{g}}$ the subalgebra of the elements of positive degree, and Z the one-dimensional subspace in $\widehat{\mathfrak{g}}$, generated by the central element t. The degree of t has been set to 0, hence $\widehat{\mathfrak{g}}_0 = \overline{\mathfrak{g}}_0 \oplus Z$. Recall that Lemma 1.3 implies that $\mathfrak{g} \subset \widehat{\mathfrak{g}}_0$ (as a Lie subalgebra).

Lemma 2.11. *A direct sum* $\widehat{\mathfrak{g}}'_+ = \widehat{\mathfrak{g}}_0 \oplus \widehat{\mathfrak{g}}_+$ *is a Lie subalgebra in* $\widehat{\mathfrak{g}}$.

Proof. Recall that by Proposition 1.2 $\mathcal{A}_0 \oplus \mathcal{A}_+$ is an (associative) subalgebra of functions which are regular at all points P_i, $i = 1, \ldots, N$. □

Let us choose and fix a Cartan subalgebra $\mathfrak{h} \subset \mathfrak{g}$, the corresponding Borel subalgebra $\mathfrak{b} \subset \mathfrak{g}$, and the corresponding upper nilpotent subalgebra $\mathfrak{n} \subset \mathfrak{g}$. Set $\overline{\mathfrak{b}}_0 = \bigoplus_{p=1}^{N} \mathfrak{b} \otimes A_{0,p} \subseteq \overline{\mathfrak{g}}_0$. It follows from the proof of Lemma 2.12 below that $\widehat{\mathfrak{b}} = \overline{\mathfrak{b}}_0 \oplus Z \oplus \widehat{\mathfrak{g}}_+$ is a Lie subalgebra in $\widehat{\mathfrak{g}}$. Consider it as an analog of the Borel subalgebra in $\widehat{\mathfrak{g}}$, and, like the conventional construction of Verma modules, define its one-dimensional representation. Let $V = \mathbb{C}v$ be a one-dimensional representation with the basis element v, $\lambda = \{\lambda_1, \ldots, \lambda_N\}$ be a set of weights (i.e., linear functions on $\mathfrak{h}$), and $\delta \in \mathbb{C}$. Let us define a mapping $\tau_{\lambda,\delta} : \widehat{\mathfrak{b}} \to \mathrm{End}(V)$ by the following relations. Set

$$\tau_{\lambda,\delta}|_{\widehat{\mathfrak{g}}_+} = 0, \quad \tau_{\lambda,\delta}(t) = \delta \cdot Id. \tag{2.5.1}$$

For $h \in \mathfrak{h}$, $n \in \mathfrak{n}$, and an arbitrary $p = 1, \ldots, N$ set

$$(hA_{0,p})v = \lambda_p(h)v, \quad (nA_{0,p})v = 0. \tag{2.5.2}$$

Lemma 2.12. *The mapping* $\tau_{\lambda,\delta}$ *is a representation of the Lie algebra* $\widehat{\mathfrak{b}}$.

Proof. The following relations are particular cases of the relations of Proposition 1.2:

$$\begin{aligned} A_{0,p}A_{0,p} &= A_{0,p} + B_p, & B_p &\in \mathcal{A}_+ \\ A_{0,p}A_{0,q} &= B_{p,q} & B_{p,q} &\in \mathcal{A}_+, \quad q \neq p\,. \end{aligned} \tag{2.5.3}$$

The cocycle defining the central extension vanishes on $\mathcal{A}_0 \oplus \mathcal{A}_+$. It implies, in particular, that $\widehat{\mathfrak{b}}$ is a Lie subalgebra, and $[\widehat{\mathfrak{b}}, \widehat{\mathfrak{b}}] \subseteq \widehat{\mathfrak{g}}_+$. Hence, according to (2.5.1), we have $\tau_{\lambda,\delta}([\widehat{\mathfrak{b}}, \widehat{\mathfrak{b}}]) = 0$. At the same time, by virtue of the one-dimensionality of V, all representation operators $\widehat{\mathfrak{b}}$ commute in that space. □

Definition 2.13. The linear space

$$\widehat{V}_{\lambda,\delta} := U(\widehat{\mathfrak{g}}) \otimes_{U(\widehat{\mathfrak{b}})} V \tag{2.5.4}$$

with the natural structure of $\widehat{\mathfrak{g}}$-module is referred to as a *Verma module* of the Lie algebra $\widehat{\mathfrak{g}}$ corresponding to the data (λ, δ). As usual, $U(\cdot)$ denotes the universal enveloping algebra of the corresponding Lie algebra. The $U(\widehat{\mathfrak{b}})$ acts on V by means of the representation $\tau_{\lambda,\delta}$. The set λ is referred to as the *highest weight*, the complex number δ as the *level*, and $v_{\lambda,\delta} = v_1 \otimes \cdots \otimes v_N$ as the *highest weight vector* of the module $V_{\lambda,\delta}$.

As in the conventional theory, the Verma module is defined by its highest weight and its level uniquely up to equivalence.

2.6 Fermion representations of Virasoro-type algebras

In this section, we define a projective representation of the Lie algebra $\mathcal{V}$, and, moreover, of the Lie algebra $\mathcal{D}^1_{\mathfrak{g}}$, in the fermion space $\mathcal{F}_\tau^{\infty/2}$ constructed above [62–64,67].

To this end, we first define a certain family of representations of $\mathcal{V}$ in $\mathcal{F}$. By Proposition 2.2 the space $\mathcal{F}$ comes from the space of sections of the vector bundle F on Σ. To define a $\mathcal{V}$-action on this space, we need a flat connection. Let us give a description of that connection in the analytic form discovered in [33,35], which does not explicitly appeal to vector bundles. Here, we need this description for $\mathfrak{g} = \mathfrak{gl}(n)$ only. For the other classical Lie algebras, we delay the description until Section 4.3. We would like to stress that the connections we consider here are precisely the same as those used in Section 4.3 for constructing central extensions of Lax operator algebras, and in [32] for the purpose of the theory of isomonodromic equations.

Consider a $\mathfrak{g}$-valued meromorphic 1-form ω, holomorphic outside $P_1, \dots, P_N, P_\infty$ and D, and having the following behavior at the points $\gamma \in \operatorname{supp}(D)$:

$$\omega(z) = \left(\frac{\omega_{-1}}{z} + \omega_0 + \omega_1 + \sum_{k>1} \omega_k z^k\right) dz, \tag{2.6.1}$$

where z is a local coordinate in the neighborhood of γ, and the matrix-valued coefficients of the expansion satisfy the following requirements. There exist $\tilde{\beta} \in \mathbb{C}^n$ and $\tilde{\varkappa} \in \mathbb{C}$ such that

$$\omega_{-1} = \alpha\tilde{\beta}^t, \quad \omega_0\,\alpha = \tilde{\varkappa}\alpha, \quad \tilde{\beta}^t\alpha = 1, \tag{2.6.2}$$

where α is a vector of Tyurin parameters associated with γ, and the upper t means transposition, so that $\alpha\tilde{\beta}^t$ is a rank 1 matrix, $\tilde{\beta}^t\alpha$ is a scalar.

Set

$$\nabla^{(\omega)} = d + \omega.$$

For $e \in \mathcal{V}$ set

$$\nabla^{(\omega)}_e = E(z)\left(\frac{\partial}{\partial z} + w(z)\right),$$

where z is a local coordinate, $e = E(z)\frac{\partial}{\partial z}$, $\omega = w(z)dz$ are local coordinate representations of the vector field e and the 1-form ω. We will refer to $\nabla^{(\omega)}$ as the connection given by the 1-form ω, and to $\nabla_e^{(\omega)}$ as the covariant derivative along the vector field e.

Proposition 2.14. *$\mathcal{F}$ is invariant with respect to $\nabla_e^{(\omega)}$ for any ω satisfying* (2.6.1), (2.6.2), *and any $e \in \mathcal{V}$.*

Proof. Basically we must prove that for any ψ satisfying (2.2.1) the $\nabla_e^{(\omega)}\psi$ also satisfies it.

In the neighborhood of a $\gamma \in \operatorname{supp} D$ we have

$$\frac{\partial}{\partial z}\psi = -\frac{\lambda\alpha}{z^2} + \psi_1 + \cdots,$$

$$w(z)\left(\frac{\lambda\alpha}{z} + \psi_0 + \cdots\right) = \frac{\lambda\alpha(\tilde{\beta}^t\alpha)}{z^2} + \frac{\omega_0\alpha + \alpha(\tilde{\beta}^t\psi_0)}{z} + \cdots.$$

By virtue of the relations $\omega_0\,\alpha = \tilde{\varkappa}\alpha$, $\tilde{\beta}^t\alpha = 1$ we obtain

$$\left(\frac{\partial}{\partial z} + w(z)\right)\psi = \frac{\tilde{\lambda}\alpha}{z} + \tilde{\psi}_0 + \cdots,$$

where $\tilde{\lambda} = \tilde{\varkappa} + \tilde{\beta}^t\psi_0$, $\tilde{\psi}_0 = \psi_1$. The right-hand side of the last relation is also of the form (2.2.1). The multiplication by $E(z)$ will not spoil this form, because E is scalar, and holomorphic at the point $z = 0$ (equivalently, e is holomorphic at γ). □

It is easily verified that for any $\psi \in \mathcal{F}$, $g \in \mathcal{A}$, $e, f \in \mathcal{L}$

$$\nabla_e^{(\omega)}(g\cdot\psi) = (eg)\cdot\psi + g\cdot\nabla_e^{(\omega)}\psi, \qquad \nabla_{g\cdot e}^{(\omega)}\psi = g\cdot\nabla_e^{(\omega)}\psi, \tag{2.6.3}$$

and

$$\nabla_{[e,f]}^{(\omega)} = \left[\nabla_e^{(\omega)}, \nabla_f^{(\omega)}\right]. \tag{2.6.4}$$

The first two relations are simply a conventional definition of a connection. The last relation expresses the property of flatness of the connection $\nabla^{(\omega)}$. It is equivalent to the statement that the mapping $e \to \nabla_e^{(\omega)}$ gives a representation of the Lie algebra $\mathcal{V}$ in the vector space $\mathcal{F}$. Then the first of the relations (2.6.3) is equivalent to the fact that this representation continues to $\mathcal{D}^1$ (indeed, it can also be written as $\nabla_e^{(\omega)}(g\cdot\psi) - g\cdot\nabla_e^{(\omega)}\psi = (eg)\cdot\psi$, and further on $[\nabla_e^{(\omega)}, g] = [e, g]$).

Let us define the corresponding action of the Lie algebra $\mathcal{V}$ in $\mathcal{F}_\tau$ as

$$e(\psi\otimes v) = \nabla_e\psi\otimes v \quad \text{for any} \quad e\in\mathcal{V},\ \psi\in\mathcal{F},\ v\in V_\tau. \tag{2.6.5}$$

It can be directly checked that (2.3.5) and (2.6.5) together give a representation of the Lie algebra $\mathcal{D}^1_{\mathfrak{g}}$ in the space $\mathcal{F}_\tau$.

Lemma 2.15 ([67]). *With respect to the index given by* (2.4.1), *the* $\mathcal{D}^1_{\mathfrak{g}}$*-module* $\mathcal{F}$ *is almost-graded.*

Proof. We keep the notation and setup of Section 2.4 here. In particular, for simplicity we consider the two-point situation. Let us also assume (again for simplicity) that ω has logarithmic singularities at the points $P_\pm$. Then $\partial/\partial z$ and multiplying by w equally reduce ord ψ at $P_\pm$, subtracting 1 from it, except if deg $\psi = 0$. By definition of the basis vector fields (1.3.9), the order $e_m\psi_{n,j}$ is equal to $n+m$ at P_+, and to $-n-m-3g$ at P_- for $n \neq 0$; for $n = 0$ these orders are equal to $n+m$ and $-n-m-3g$, respectively. Hence, for some constants $D^{k,j'}_{m,n,j}$ the following relations hold:

$$e_m\psi^i_{n,j} = \sum_{k=m+n+\varepsilon}^{m+n+3g+\varepsilon}\sum_{j'=0}^{l-1} D^{k,j'}_{m,n,j}\psi^i_{k,j'}, \tag{2.6.6}$$

where $\varepsilon = 0$ for $n \neq 0$, and $\varepsilon = 1$ for $n = 0$. □

By virtue of almost-grading, the construction of the vector field action in $\mathcal{F}^{\infty/2}$ and $\mathcal{F}^{\infty/2}_\tau$ is quite the same as in the affine Lie algebra case (Section 2.4), but there is no need to repeat it here.

Let us prove the following helpful assertion: let $\mathcal{V}^{(2)}_+ \subset \mathcal{V}$ be the subalgebra consisting of elements vanishing with the order at least two at P_+.

Lemma 2.16. *The elements of* $\mathcal{V}^{(2)}_+$ *annihilate the vacuum vectors of fermion representations.*

Proof. By definition (Section 1.3), $e_m \in \mathcal{V}^{(2)}_+$ if and only if $m \geq 1$. In this case, according to (2.6.6), the action of the vector field e_m increases value of the index n: $k > n$ for all $\psi^i_{k,j'}$ on the right-hand side of the relation (2.6.6). As soon as a vacuum monomial contains $\psi^i_{n,j}$, it also contains *all* $\psi^i_{k,j'}$, $k > n$ to the right of $\psi^i_{n,j}$. □

The joint action of $\mathcal{V}$ and $\overline{\mathfrak{g}}$ in $\mathcal{F}^{\infty/2}$ turns it into a projective module over the Lie algebra $\mathcal{D}^1_{\mathfrak{g}} = \mathcal{V} \oplus \overline{\mathfrak{g}}$. The cocycle of this module is local for the same obvious reason as in Section 2.4.

We return now to the proof of Theorem 2.9. Recall that this theorem claims that *if* $\mathfrak{g}$ *is simple, then the cocycle of the fermion representation of the Lie algebra* $\overline{\mathfrak{g}}$ *has the form*

$$\widehat{\gamma}(xA, yB) = \lambda \cdot (x, y)\operatorname{res}_{P_+} A dB, \tag{2.6.7}$$

where (x, y) *is the Killing form of* $\mathfrak{g}$, $\lambda \in \mathbb{C}$.

Proof of Theorem 2.9. According to Lemma 2.10

$$\widehat{\gamma}(xA, yB) = \alpha(x, y)\,\gamma(A, B), \tag{2.6.8}$$

where α is an invariant symmetric bilinear form on $\mathfrak{g}$, γ is a cocycle on $\mathcal{A}$. Since $\mathfrak{g}$ is simple, we immediately obtain that $\alpha(x, y) = \lambda_1 \cdot (x, y)$ where $\lambda_1 \in \mathbb{C}$, $\lambda_1 \neq 0$.

Since the $\overline{\mathfrak{g}}$-action in $\mathcal{F}^{\infty/2}$ is a restriction of the $\mathcal{D}^1_{\mathfrak{g}}$-action, the cocycle $\widehat{\gamma}$ is a restriction of a cocycle on $\mathcal{D}^1_{\mathfrak{g}}$. We will keep the same notation $\widehat{\gamma}$ for the latter.

By the definition of cocycle, for the triple of Lie algebra $\mathcal{D}^1_{\mathfrak{g}}$ elements e, xA, yB we have

$$\widehat{\gamma}(e, [xA, yB]) + \widehat{\gamma}(yB, [e, xA]) + \widehat{\gamma}(xA, [yB, e]) = 0. \tag{2.6.9}$$

Let $x, y \in \mathfrak{g}$ belong to the same Cartan subalgebra, hence commute, and $(x, y) \neq 0$. Then $[xA, yB] = 0$ (in $\overline{\mathfrak{g}}$), hence (2.6.9) together with (2.6.8) gives

$$(y, x)\gamma(B, eA) - (x, y)\gamma(A, eB) = 0. \tag{2.6.10}$$

Dividing by (x, y) we obtain

$$\gamma(B, eA) = \gamma(A, eB), \tag{2.6.11}$$

i.e., the $\mathcal{V}$-invariance of the cocycle γ (cf. (1.5.7)). By Theorem 1.8(a), $\gamma(A, B) = \lambda_2 \operatorname{res}_{P_+} AdB$, $\lambda_2 \in \mathbb{C}$. □

2.7 Sugawara representation

Let $\mathfrak{g}$ be a finite-dimensional reductive Lie algebra over $\mathbb{C}$. Let us fix a nondegenerate invariant symmetric bilinear form α on $\mathfrak{g}$, hence the cocycle γ given by (1.6.2). Let $\widehat{\mathfrak{g}}$ be the corresponding affine Krichever–Novikov algebra introduced in Section 1.6 (see (1.6.1) and (1.6.2)). In this section, we prefer to speak of $\widehat{\mathfrak{g}}$-modules rather than of projective $\overline{\mathfrak{g}}$-modules. A representation of $\widehat{\mathfrak{g}}$ is called admissible if its restriction to $\overline{\mathfrak{g}}$ is admissible (see page 20 for the definition). Here, we consider admissible $\widehat{\mathfrak{g}}$-modules such that the central element t acts by a scalar operator. If that operator is equal to cI ($c \in \mathbb{C}$, I is the unit operator) then c is referred to as a *level* of the module (representation).

If the Lie algebra $\mathfrak{g}$ is commutative or simple, then for any admissible $\widehat{\mathfrak{g}}$-module of noncritical level (i.e., nonzero in the commutative case and different from a certain value defined below in the simple case) the Sugawara construction gives a projective representation of $\mathcal{V}$ in the same space. This representation is referred to as *Sugawara representation*. The construction in question goes back to H. Sugawara. For loop algebras, it was developed by M. Wakimoto, V. Kac, and I. Frenkel (see [15, 17] and references therein). For commutative Krichever–Novikov algebras in the two-point case, the Sugawara construction is given in [27]. The noncommutative case was first considered in the physics literature [4]. A mathematically consistent treatment is given in our work [51] where also the multi-point version of the construction is given.

In the case of an arbitrary complex reductive Lie algebra $\mathfrak{g}$ the Sugawara representation is defined as the sum of the Sugawara representations corresponding to its simple and commutative ideals (Theorem 2.23).

Let us turn to the presentation of the Sugawara construction. Assume that V is an arbitrary admissible representation of level c of the Lie algebra $\widehat{\mathfrak{g}}$. For any $u \in \mathfrak{g}$, $A \in \mathcal{A}$ denotes by $u(A)$ the operator in V corresponding to the element $u \otimes A$. The operator $u(A_{n,p})$ is also denoted by $u(n,p)$. Fix a basis u_i, $i = 1, \dots, \dim \mathfrak{g}$ in the Lie algebra $\mathfrak{g}$, and a dual basis u^i, $i = 1, \dots, \dim \mathfrak{g}$ with respect to the form α. The *Casimir element* $\Omega^0 = \sum_{i=1}^{\dim \mathfrak{g}} u_i u^i$ of the universal enveloping algebra $U(\mathfrak{g})$ does not depend on this choice. To simplify notation, we write $u(n,p)u(m,q)$ instead of $\sum_i u_i(n,p)u^i(m,q)$.

Let us define the *generating Sugawara operator* referred to as the *energy-momentum tensor* in the physics literature:

$$T(P) := \frac{1}{2} \sum_{n,m} \sum_{p,s} {:}u(n,p)u(m,s){:}\, \omega^{n,p}(P)\omega^{m,s}(P) \,. \tag{2.7.1}$$

By $::$ we denote some *normal ordering*. In this section, the summation indices n, m range over $\mathbb{Z}$, and p, s range over $\{1, \dots, N\}$. Any precise form of the normal ordering is of no importance here. For example, we can take the following "standard" normal ordering ($x, y \in \mathfrak{g}$):

$$ {:}x(n,p)y(m,r){:} = \begin{cases} x(n,p)y(m,r), & n \le m \\ y(m,r)x(n,p), & n > m. \end{cases} \tag{2.7.2}$$

The expression $T = T(P)$ may be considered as a formal series of quadratic differentials with operator-valued coefficients. Taking its expansion over the basis $\Omega^{k,r}$ of quadratic differentials, obtain

$$T = \sum_k \sum_r L_{k,r} \cdot \Omega^{k,r}, \tag{2.7.3}$$

where

$$\begin{aligned} L_{k,r} &= \frac{1}{2\pi \mathrm{i}} \int_{C_S} T \cdot e_{k,r} = \frac{1}{2} \sum_{n,m} \sum_{p,s} {:}u(n,p)u(m,s){:}\, l_{(k,r)}^{(n,p)(m,s)}, \\ l_{(k,r)}^{(n,p)(m,s)} &= \frac{1}{2\pi \mathrm{i}} \int_{C_S} \omega^{n,p} \omega^{m,s} e_{k,r} \,. \end{aligned} \tag{2.7.4}$$

Formally, $L_{k,r}$ is an infinite double sum. However, for given k and m, the coefficient $l_{(k,r)}^{(n,p)(m,s)}$ turns out to be nonzero only for the finite set of values of n. It can be checked by considering the residues of integrands. We still have an infinite sum over m. Let us take an arbitrary $v \in V$. In virtue of normal ordering and the admissibility of the representation, only a finite number of summands acts nontrivially on v. Hence the action of this infinite sum of operators on every element of V is well defined. The operators $L_{k,r}$, as well as their normalized version $L^*_{k,r}$ defined below are referred to as *Sugawara operators*.

The following two theorems contain the principal results on the Sugawara construction.

Theorem 2.17 ([51]). *Let $\mathfrak{g}$ be a finite-dimensional, commutative, or simple, Lie algebra, $2\varkappa$ the eigenvalue of its Casimir operator in the adjoint representation, and V an admissible almost-graded $\widehat{\mathfrak{g}}$-module of level c. If $c+\varkappa \neq 0$, then the normalized "modes"*

$$L^*_{k,r} = \frac{-1}{2(c+\varkappa)} \sum_{n,m} \sum_{p,s} {:}u(n,p)u(m,s){:}\, l^{(n,p)(m,s)}_{(k,r)} \tag{2.7.5}$$

of the energy-momentum tensor give well-defined operators on V and define an admissible representation of $\mathcal{V}$ with the cocycle

$$\chi(e,f) = \frac{c \cdot \dim \mathfrak{g}}{c+\varkappa} \cdot \frac{1}{24\pi i} \int_{C_S} \left(\frac{1}{2}(e'''f - ef''') - R \cdot (e'f - ef') \right) dz, \tag{2.7.6}$$

where R is an appropriate meromorphic projective connection with poles only at the points $P_\pm$, $\operatorname{ord}_{P_+} R \geq -1$, $\operatorname{ord}_{P_-} R \geq -2$.

Remark 2.18. Both the level c, and half the eigenvalue of the Casimir operator $\varkappa$ depend on the normalization of the bilinear form α. In both cases the dependence is inversely proportional, so that condition $c+\varkappa \neq 0$ is invariant. If α is normalized so that the longest roots have the square length 2 (with respect to α) then $\varkappa$ is an integer referred to as the *dual Coxeter number* [17, Remark 10.2]. We do not fix any normalization of α here.

Theorem 2.19 ([51]). *The following commutation relations take place:*

$$\left[L_{k,r}, x(m,s)\right] = -(c+\varkappa)\, x(e_{k,r} A_{m,s}),$$

where eA is a derivative of A along the vector field e.

We prove Theorems 2.17 and 2.19 in Section 2.8. Here, we continue establishing the basic properties of the Sugawara construction.

Proposition 2.20. *The action of the Sugawara operators in V is almost-graded.*

Proof. We must prove that there are constants M_1, M_2, such that for any fixed homogeneous element $\psi_s \in V$ of degree s, and any k, the following relation holds:

$$k + s + M_1 \leq \deg(L^*_{k,r}\psi_s) \leq k + s + M_2. \tag{2.7.7}$$

Let us consider the coefficients $l_{(k,r)}^{(n,p)(m,s)}$ given by the integrals (2.7.4). Such an integral can be nonzero only if the integrand has poles at the points in I and at P_∞. Making use of the explicit formulae (1.3.2) and (1.3.3) for the orders at those points, we obtain

$$k \leq n + m \leq k + C(g, N) \tag{2.7.8}$$

with a rational constant $C(g, N) \geq 0$, depending on the genus g, and the number of marked points N. By virtue of the almost-grading of V being regarded as a $\widehat{\mathfrak{g}}$-module, there exist constants c_1, c_2, such that for all m, r, s

$$m + s + c_1 \leq \deg\big(u(m, r)\psi_s\big) \leq m + s + c_2 \tag{2.7.9}$$

if $u(m, r)\psi_s \neq 0$. Hence,

$$n + m + s + 2c_1 \leq \deg\big(:u(n, p)u(m, r): \psi_s\big) \leq n + m + s + 2c_2 \tag{2.7.10}$$

if $:u(n, p)u(m, r): \psi_s \neq 0$. Making use of (2.7.8), we obtain the relation (2.7.7) if set $M_1 = 2c_1$ and $M_2 = 2c_2 + C(g, N)$. □

For $e = \sum_{n,p} a_{n,p} e_{n,p} \in \mathcal{V}$ $(a_{n,p} \in \mathbb{C})$ set $T(e) = \sum_{n,p} a_{n,p} L_{n,p}^*$. Then $T(e) \in \operatorname{End} V$. By virtue of the Krichever–Novikov duality the Sugawara operator $T(e)$ corresponding to a vector field $e \in \mathcal{V}$ can be equivalently defined as

$$T(e) = \frac{-1}{c + \varkappa} \cdot \frac{1}{2\pi \mathrm{i}} \int_{C_S} T(P)e(P). \tag{2.7.11}$$

Summing up Theorems 2.17 and 2.19 and Proposition 2.20, we obtain the following theorem.

Theorem 2.21. *Let $\mathfrak{g}$ be a commutative or simple Lie algebra. The mapping of the Lie algebra $\mathcal{D}_{\mathfrak{g}}^1$ to End(V) given by the formulae $e \to T(e)$, $uA \to u(A)$ is an almost-graded projective representation of the Lie algebra $\mathcal{D}_{\mathfrak{g}}^1$ having zero mixing cocycle. In particular, the following relations hold:*

(i) $[T(e), T(f)] = T([e, f]) + \chi(e, f)$, *where χ is a cocycle on $\mathcal{V}$, given by (2.7.6);*
(ii) $[T(e), u(A)]=u(eA)$.

A projective representation T of the Lie algebra $\mathcal{V}$ is referred to as the *Sugawara representation* corresponding to the given admissible representation u of the Lie algebra $\widehat{\mathfrak{g}}$.

Theorem 2.21 generalizes to the case of reductive Lie algebras. Let $\mathfrak{g}$ be a reductive Lie algebra having the decomposition (1.6.6). The elements $x \in \mathfrak{g}$ can be represented in the form $x = \sum_{i=0}^n x_i$ where $x_i \in \mathfrak{g}_i$.

Lemma 2.22 ([49], Lemma 3.11). *For an arbitrary cocycle γ on $\overline{\mathfrak{g}}$, and any $A, B \in \mathcal{A}$, $x, y \in \mathfrak{g}$*

$$\gamma(x_i A, y_j B) = 0 \textit{ for } i \neq j$$

(assuming that there is a decomposition for y similar to the above decomposition for x).

Proof. By the definition of cocycle, for any $x, z \in \mathfrak{g}_i$, $y \in \mathfrak{g}_j$ we have

$$\gamma\big([xA, yB], z\big) + \gamma\big([z, xA], yB\big) + \gamma\big([yB, z], xA\big) = 0.$$

By the relation $[\mathfrak{g}_i, \mathfrak{g}_j] = 0$ the first and the third summands vanish. This implies $\gamma([z,x]A, yB) = 0$. By the simplicity of the Lie algebra $\mathfrak{g}_i$ we have $[\mathfrak{g}_i, \mathfrak{g}_i] = \mathfrak{g}_i$. Hence for any $x_i \in \mathfrak{g}_i$, $y_j \in \mathfrak{g}_j$ we can take x, z such that $[z, x] = x_j$, $y = y_j$, which proves our statement. □

Lemma 2.22 implies that

$$\gamma(xA, yB) = \sum_{i=0}^{M} \gamma(x_i A, y_i B). \tag{2.7.12}$$

Given an admissible representation of the Lie algebra $\widehat{\mathfrak{g}}$ in the space V, we will obtain the representations of its ideals $\widehat{\mathfrak{g}}_i$ in the same space. For these, we can define (normalized) Sugawara operators $T_i(e)$, $i = 0, 1, \ldots, M$. We set by definition that

$$T(e) = \sum_{i=0}^{M} T_i(e). \tag{2.7.13}$$

Theorem 2.23. *The statements of Theorem 2.21 still hold if $\mathfrak{g}$ is a reductive Lie algebra. In this case the cocycle of the representation T is equal to the sum of the cocycles of the representations T_i, $i = 0, 1, \ldots, M$.*

Proof. For $x, y \in \mathfrak{g}$, let x_i, y_j define their components as above. Then $[x_i(A), y_j(B)] = [x_i, y_j](AB) + \gamma(x_i A, y_j B) \cdot id$. If $i \neq j$, then the elements x_i and y_j commute and the cocycle vanishes. Hence, the elements $x_i(A)$, $y_j(B)$ commute in $\widehat{\mathfrak{g}}$. Let T_k be the Sugawara representation corresponding to the representation of $\widehat{\mathfrak{g}}_k$ in V. Since the operators of the representation T_i express via the operators $x_i(A)$, and the operators of the T_j via $x_j(A)$, the representations T_i and T_j commute for $i \neq j$. Hence

$$\left[\sum_{i=0}^{M} T_i(e), \sum_{j=0}^{M} T_j(f)\right] = \sum_{i=0}^{M} [T_i(e), T_i(f)].$$

This implies both that T is a projective representation and that the mixing cocycle is trivial.

Further on, let us prove the assertion (ii) of Theorem 2.21 for a reductive Lie algebra. For $x_i \in \mathfrak{g}_i$, $A \in \mathcal{A}$, $e \in \mathcal{V}$ we have $[T_j(e), x_i(A)] = 0$, as above. By Theorem 2.21 the following relation holds:

$$\left[T_k(e), x_k(A)\right] = x_k(eA), \quad k = 0, 1, \ldots, M,$$

implying that

$$[T(e), x(A)] = \left[\sum_{k=0}^{M} T_k(e), \sum_{i=0}^{M} x_i(A)\right] = \sum_{i=0}^{M} x_i(eA) = x(eA). \qquad \square$$

In the particular case where the representation V of the Lie algebra $\widehat{\mathfrak{g}}$ is generated by a vacuum, the following assertion is instructive. We formulate it in the case of two marked points, i.e., the index p is omitted in the notation of the basis elements and the corresponding Sugawara operators. Let $\mathcal{L}_+^{(2)} \subset \mathcal{V}$ be the Lie subalgebra of vector fields vanishing at P_+ with the order at least two.

Lemma 2.24. *The Sugawara operators of the elements of $\mathcal{L}_+^{(2)}$ annihilate the vacuum vector of the representation of the Lie algebra $\widehat{\mathfrak{g}}$, to which they correspond.*

Proof. By definition, $e_k \in \mathcal{L}_+^{(2)}$ if, and only if $k \geq 1$. Consider the Sugawara operators L_k for $k \geq 1$. We want to show that if $l_k^{mn} \neq 0$ then either $m \geq 1$ or $n \geq 1$. If it is true then, taking account of the normal ordering, the second factor of the product $:u(m)u(n):$ is the representation operator of an element of the subalgebra $\widehat{\mathfrak{g}}_+$. Hence, this product annihilates the vacuum.

Consider the expression for l_k^{mn} in (2.7.4). By Section 1.3

$$\operatorname{ord}_{P_+} \omega^m = -m - 1, \ \operatorname{ord}_{P_+} \omega^n = -n - 1, \ \operatorname{ord}_{P_+} e_k = k + 1.$$

Thus $\operatorname{ord}_{P_+} \omega^m \omega^n e_k = -m - n + k - 1$. For the residue of the 1-form $\omega^m \omega^n e_k$ at the point P_+ to be nontrivial, the 1-form must have a pole at P_+. Hence $-m - n + k - 1 \leq -1$. It implies that $m + n \geq k \geq 1$, hence either $m > 0$ or $n > 0$. Then the operator $:u(n)u(m):$ near that l_k^{mn} in the first of the relations (2.7.4), annihilates the vacuum. $\square$

2.8 Proof of the main theorems for the Sugawara construction

This section may be skipped for the first reading. A simplified version of the proofs (for the loop and Virasoro algebras) can be found in [17]. We actually follow the lines of that proof, even though this is not obvious.

To simplify the notation, we give the proofs in the case of two marked points. Then the index of a marked point disappears. In principle, adding this index changes nothing in the proofs. A full treatment of the multipoint case will be found in the forthcoming book [50] by M. Schlichenmaier.

2.8.1 Main theorems in the form of relations with structure constants

The following statement is a reformulation of Theorem 2.19 via structure constants. It is a key ingredient of the construction.

Proposition 2.25. *Let $\mathfrak{g}$ be a simple or commutative Lie algebra, then*

$$[L_k, x(r)] = -(c+\varkappa)\sum_v K^v_{r,k}\, x(v) \tag{2.8.1}$$

where $$K^v_{r,k} = \frac{1}{2\pi\,\mathrm{i}}\oint_C \omega^v(Q)e_k(Q)dA_r(Q) = \sum_m l_k^{vm}\gamma_{mr}, \tag{2.8.2}$$

$$\gamma_{mr} = \gamma(A_m, A_r) = \frac{1}{2\pi\,\mathrm{i}}\oint_C A_m(Q)dA_r(Q). \tag{2.8.3}$$

The result does not depend on a normal ordering.

Observe that all the sums here are indeed finite, thus well defined. In the case of loop and Virasoro algebras, the relation (2.8.1) can be written as follows: $[L_k, x(r)] = -(c+\varkappa)r\, x(r+k)$ (see [17, Prop. 10.1]). Let us delay its proof until page 44. Here we want to prove (2.8.2). Let us define the *delta distribution* [26]:

$$\Delta(Q', Q) = \sum_{m=-\infty}^{\infty} A_m(Q')\omega^m(Q).$$

This is a formal series, but for any basis elements A_m, ω^n the following integrals are well defined, and, moreover, easily calculated using the duality relations:

$$\int_C \Delta(Q', Q)A_m(Q) = A_m(Q'), \quad \int_{C'} \Delta(Q', Q)\omega^n(Q') = \omega^n(Q).$$

Proof of the relation (2.8.2). Making use of the delta-distribution, we obtain

$$\begin{aligned}\sum_m l_k^{vm}\gamma_{mr} &= \sum_m \frac{1}{2\pi\,\mathrm{i}}\oint_C \omega^v(Q)\omega^m(Q)e_k(Q)\int_{C'} A_m(Q')dA_r(Q') \\ &= \frac{1}{(2\pi\,\mathrm{i})^2}\iint\limits_{C\,C'} \omega^v(Q)e_k(Q)dA_r(Q')\Delta(Q', Q) \\ &= \int_{C'} \omega^v(Q')e_k(Q')dA_r(Q') = K^v_{r,k}.\end{aligned}$$

□

Proposition 2.26. $[L_k, x(r)] = -(c+\varkappa)\, x(\nabla_{e_k} A_r)$.

Proof. In local coordinates, $\nabla_{e_k} A_r$ are written as $e_k(z)\frac{dA_r(z)}{dz}$, and hence

$$(\nabla_{e_k} A_r)(Q) = e_k(Q)dA_r(Q) = \sum_v \beta^v_{r,k} A_v(Q).$$

By duality, we have for the coefficients $\beta^v_{r,k}$

$$\beta^v_{r,k} = \frac{1}{2\pi\,\mathrm{i}} \oint_C e_k(Q)dA_r(Q)\omega^v(Q) = K^v_{r,k}.$$

Hence our assertion follows from Proposition 2.25. □

Proposition 2.27. *The operators $L_k \in \mathfrak{gl}(V)$ and $id \in \mathfrak{gl}(V)$ form a Lie subalgebra in $\mathfrak{gl}(V)$ with the commutation relations*

$$[L_k, L_l] = -(c+\varkappa)\sum_n C^n_{kl} L_n - \frac{1}{2}c(c+\varkappa)\dim\mathfrak{g}\cdot\chi_{kl}\cdot id, \tag{2.8.4}$$

where C^n_{kl} are structure constants of the Lie algebra of vector fields $\mathcal{V}$, and

$$\chi_{kl} = \psi_{kl} + \widehat{\chi}_{kl}, \quad \psi_{kl} = \sum_{s,v}\sum_{n=0}^{v+1} C^s_{kl} l^{nv}_s \gamma_{nv} = \sum_v \sum_{n=0}^{v+1} E^{nv}_{kl}\gamma_{nv}, \tag{2.8.5}$$

$$E^{nv}_{kl} = \frac{1}{2\pi\,\mathrm{i}}\oint_C [e_k, e_l]\cdot\omega^n\omega^v, \qquad \widehat{\chi}_{kl} = \Bigg(\sum_{\substack{n>0\\ v\le 0}} - \sum_{\substack{n\le 0\\ v>0}}\Bigg) K^n_{v,k} K^v_{n,l}. \tag{2.8.6}$$

The quantities χ_{kl} can be nonzero only if $-6g \le k+l \le 0$. Another normal ordering would change the summation range over n, v in the definition of ψ_{kl}, and hence it would change the values χ_{kl}. At the upper limit we obtain

$$\chi_{k,-k} = -\frac{1}{6}(k^3 - k)\,. \tag{2.8.7}$$

The proof of this assertion is also delayed until the end of the section. Here we will derive Theorem 2.17 from it. For this purpose we need one more simple lemma.

Lemma 2.28 ([27], relation (3.14)). *Let χ_R be the cocycle of the form* (1.5.8), *R be a meromorphic projective connection with poles at the points $P_\pm$. Then*

(a) *vanishing the $\chi_R(e_k, e_l)$ for $0 < k+l < -6g$ is equivalent to the fact that R has at most double poles;*

(b) *let $R(z_+) = \alpha_+ z_+^{-2}(1 + O(z_+))$ be the local form of the projective connection at the point P_+, then*

$$\chi_R(e_k, e_{-k}) = \frac{1}{12}(k^3 - k - 2\alpha_+ k).$$

The proof comes to a calculation of residues.

Proof of Theorem 2.17. If $c + \varkappa \neq 0$ then it is possible to take normalized elements $L_k^* = \frac{-1}{c+\varkappa} L_k$ and obtain

$$[L_k^*, L_l^*] = \sum_n C_{kl}^n L_n^* - \frac{c}{2(c+\varkappa)} \dim \mathfrak{g} \cdot \chi_{kl} \cdot id. \tag{2.8.8}$$

The Jacoby identity in $\mathfrak{gl}(V)$ implies that χ_{kl} defines a two-cocycle χ for the Lie algebra $\mathcal{V}$. Hence we have obtained a projective representation of the algebra $\mathcal{V}$. According to Proposition 2.27, the cocycle χ_{kl} is local. Let us show that its cohomology class is nontrivial.

Every 2-coboundary, by definition, is of the form $\phi([e, f])$, where $e, f \in \mathcal{V}$, ϕ is a linear function on $\mathcal{V}$. By virtue of the duality relations (1.3.1) and the notation (1.3.5), the quadratic differentials Ω^n form a basis which is dual for the vector field basis e_m. Hence every linear function ϕ on $\mathcal{V}$ is their combination, perhaps infinite, i.e., is given by a quadratic differential Ω as

$$\phi(e) = \operatorname{res}_{P_+} \Omega e. \tag{2.8.9}$$

Assume that the cocycle χ_{kl} has such a form. Then for a pair of vector fields e_k, e_{-k}, we obtain

$$\chi_{k,-k} = \operatorname{res}_{P_+} w(E_k' E_{-k} - E_k E_{-k}'), \tag{2.8.10}$$

where $\Omega = w dz^2$, $e_k = E_k \partial_z$, $e_{-k} = E_{-k}\partial_z$ are expressions of the corresponding objects in local coordinates in the neighborhood of P_+. Since $E_k = z^{k+1}\partial_z$, $E_{-k} = z^{-k+1}\partial_z$, we have $E_k' E_{-k} - E_k E_{-k}' = 2kz$. This would mean that the residue in (2.8.10) depends linearly on k, which contradicts the relation (2.8.7) of Proposition 2.27, where the dependence is cubic.

By Theorem 1.8, all local cohomologically nontrivial cocycles are proportional to geometric cocycles (1.5.8) with appropriate projective connections. From Proposition 2.27, and Lemma 2.28 we conclude that the projective connection R corresponding to the cocycle χ has at most double poles at the marked points. Comparison with (2.8.7) and assertion (b) of Lemma 2.28 makes it possible to find the normalizing constant (equal to -2).

This proves Theorem 2.17. □

Remark 2.29. The expressions ψ_{kl} in (2.8.5) define the coboundary in the sense of Lie algebra cohomology. Define a linear function $\phi : \mathcal{V} \to \mathbb{C}$ by setting $\phi(e_s) = \sum_v \sum_{n=0}^{v+1} l_s^{nv} \gamma_{nv}$ (this sum is finite), then

$$\psi_{kl} = \phi\Big(\sum_s C_{k,l}^s e_s\Big) = \phi([e_k, e_l]).$$

Another normal ordering would result in a different summation range in the above definition of ϕ. It shows again that the cohomology class of χ_{kl} (= equivalence class of the corresponding central extension) is invariant with respect to the choice of a normal ordering.

2.8.2 End of the proof of the main theorems

First, let us prove the following lemma (we do not assume any summation over repeating indices here).

Lemma 2.30. *As above, let $\{u_i\}$, $\{u^i\}$ be a pair of dual bases in $\mathfrak{g}$, and $\Omega^0 = u_i u^i$ be a Casimir element. Then*

(1) $[\Omega^0, \mathfrak{g}] = 0$;

(2) $\sum_i [u_i, u^i] = 0$;

(3) $\sum_i [u_i A_n, u^i A_m] = -\dim \mathfrak{g} \cdot \gamma(A_n, A_m) \cdot t$;

(4) *if $\mathfrak{g}$ is a commutative or simple Lie algebra then there exists a constant $\varkappa$ such that $\sum_i ad_{u_i} \circ ad_{u^i} = 2\varkappa$.*

Proof. The assertions (1), (2), and (4) are conventional [17]. To prove (3), let us take the definition of commutator in the central extension

$$[u_i A_n, u^i A_m] = [u_i, u^i](A_n A_m) - \gamma(A_n, A_m) \cdot (u_i, u^i) \cdot t \, .$$

Here and below in this section let (x, y) denote the same as $\alpha(x, y)$. After summation over i, the first summand disappears by (2), hence we obtain the desired result. □

Further on, let us introduce one more tool. There are infinite formal sums in the definition of the operators L_k. To make them well defined, following [17], let us make use of the *cut-off procedure*. Let ψ be the function on $\mathbb{R}$ given as

$$\psi(x) = 1 \quad \text{if } |x| \le 1 \qquad \text{and} \qquad \psi(x) = 0 \quad \text{if } |x| > 1 \, . \tag{2.8.11}$$

For $\epsilon \in \mathbb{R}$ define

$$L_k(\epsilon) = \frac{1}{2} \sum_{n,m} \sum_i :u_i(n) u^i(m): \, l_k^{nm} \psi(\epsilon n). \tag{2.8.12}$$

For a given k, and any fixed n there are only finitely many values of m such that $l_k^{nm} \neq 0$. Hence for $\epsilon > 0$ the sum consists of only finitely many summands. If $v \in V$, then due to the normal ordering only finitely many operators $l_k^{nm} :u_i(n) u^i(m):$ act nontrivially on v. Hence, if we take $\epsilon > 0$ to be sufficiently small, we obtain $L_k(\epsilon) v = L_k v$. This is what we mean by $\lim_{\epsilon \to 0} L_k(\epsilon) = L_k$.

If we throw the normal ordering symbol in (2.8.12) off, we will obtain $\widetilde{L}_k(\epsilon)$ which is well-defined for $\epsilon \neq 0$. For any not normally ordered pair (n,m) let us consider the expression $\sum_i [u_i(n), u^i(m)]$ which is a scalar operator by Lemma 2.30 (3), hence $L_k(\epsilon) = \widetilde{L}_k(\epsilon) + \alpha \cdot t$ where α is a scalar if $\epsilon \neq 0$. In particular, computing commutators we can forget of the normal ordering while $\epsilon \neq 0$.

Proof of Proposition 2.25. By definition, and by the Leibniz formula for commutators we have

$$\begin{aligned} R_\epsilon := 2\Big[\widetilde{L}_k(\epsilon), x(r)\Big] &= \sum_{n,m}\sum_i \big[u_i(n)u^i(m), x(r)\big]\, l_k^{nm}\psi(\epsilon n) \\ &= \sum_{n,m}\sum_i \Big(u_i(n)\big[u^i(m), x(r)\big] + \big[u_i(n), x(r)\big]u^i(m)\Big)\, l_k^{nm}\psi(\epsilon n)\,. \end{aligned} \tag{2.8.13}$$

Further, every commutator writes as

$$[u^i(m), x(r)] = [u^i, x](A_m A_r) - (u^i, x)\gamma_{mr} \cdot c$$

(observe that $tv = c \cdot v$). Hence

$$R_\epsilon = A_\epsilon + B_\epsilon - (C_\epsilon + D_\epsilon), \tag{2.8.14}$$

where

$$A_\epsilon = \sum_{n,m}\sum_i u_i(n)[u^i, x](A_m A_r) l_k^{nm}\psi(\epsilon n),$$

$$B_\epsilon = \sum_{n,m}\sum_i [u_i, x](A_n A_r) u^i(m) l_k^{nm}\psi(\epsilon n),$$

$$C_\epsilon = \sum_{n,m}\sum_i u_i(n)(u^i, x)\gamma_{mr} l_k^{nm} c\psi(\epsilon n),$$

$$D_\epsilon = \sum_{n,m}\sum_i (u_i, x) u^i(m)\gamma_{nr} l_k^{nm} c\psi(\epsilon n)\,.$$

Making use of the relation $\sum_i u_i \otimes A_n(u^i, x) = (\sum_i (u^i, x)u_i) \otimes A_n = x \otimes A_n = x(n)$, we obtain $C_\epsilon = \sum_{n,m} x(n)\gamma_{mr} l_k^{nm} c\psi(\epsilon n)$ and $D_\epsilon = \sum_{n,m} x(m)\gamma_{nr} l_k^{nm} c\psi(\epsilon n)$. There are only finitely many summands with fixed r and k. Hence, in the limit we obtain

$$\lim_{\epsilon\to 0}(C_\epsilon + D_\epsilon) = 2c \cdot \sum_n \Big(\sum_m l_k^{nm}\gamma_{mr}\Big) x(n) = 2c \sum_n K_{r,k}^n x(n)\,. \tag{2.8.15}$$

Here we made use of (2.8.2). The sums A_ϵ, B_ϵ do not converge separately as $\epsilon \to 0$ but $A_\epsilon + B_\epsilon$ does. First, substitute $A_m A_r = \sum_s \alpha_{mr}^s A_s$ where $\alpha_{mr}^s = \frac{1}{2\pi \mathrm{i}} \oint_C A_m A_r \omega^s$.

We will obtain

$$A_\epsilon = \sum_{n,m,s} \sum_i u_i(n)[u^i, x](s)\alpha^s_{mr} l^{nm}_k \psi(\epsilon n),$$
$$B_\epsilon = \sum_{n,m,s} \sum_i [u_i, x](s) u^i(m)\alpha^s_{nr} l^{nm}_k \psi(\epsilon n).$$

Further on, introduce normal ordering and the compensating commutators for those pairs which are not ordered. We obtain $A_\epsilon = A^{(1)}_\epsilon + A^{(2)}_\epsilon$, and $B_\epsilon = B^{(1)}_\epsilon + B^{(2)}_\epsilon$ where $A^{(1)}_\epsilon$ (respectively, $B^{(1)}_\epsilon$) is the same expression but containing the normal ordering symbol. A compensating commutator reads as

$$\big[u_i(n), [u^i, x](s)\big] = \big[u_i, [u^i, x]\big](A_n A_s) - \gamma_{ns}\big(u_i, [u^i, x]\big)c.$$

After summation over i, the second terms disappear by $(u_i, [u^i, x]) = ([u_i, u^i], x)$, and by Lemma 2.30 (2). For the first terms we obtain $2\varkappa \cdot x(A_n A_s) = 2\varkappa \sum_v \alpha^v_{ns} x(v)$ by Lemma 2.30 (4). Doing the same with $B^{(2)}_\epsilon$, we obtain

$$A^{(2)}_\epsilon + B^{(2)}_\epsilon = 2\varkappa \sum_v \Big(\sum_{s,m} \sum_{n>s} \alpha^v_{ns}\alpha^s_{mr} l^{nm}_k \psi(\epsilon n) - \sum_{n,m} \sum_{s>m} \alpha^v_{sm}\alpha^s_{nr} l^{nm}_k \psi(\epsilon n) \Big) x(v). \tag{2.8.16}$$

Observe that none of the sums make sense as such, as $\epsilon = 0$. Before further considering (2.8.16), let us show that $A^{(1)}_0 + B^{(1)}_0 = 0$. First, change the summation variables in the sum corresponding to $B^{(1)}_\epsilon$ as follows: $s \to n \to m \to s$. By normal ordering, $A^{(1)}_0$ and $B^{(1)}_0$ are well-defined operators in the sense of, having been applied to a $v \in V$, they will give only finitely many nonzero terms as the result. Hence we can forget about the factor ψ.

Let us define the constants

$$F^{sn}_{r,k} = \sum_m \alpha^s_{mr} l^{nm}_k = \frac{1}{2\pi \mathrm{i}} \oint_C A_r(Q)\omega^s(Q)\omega^n(Q)e_k(Q).$$

Lemma 2.31. *The constants $F^{sn}_{r,k}$ are symmetric in the upper indices: $F^{sn}_{r,k} = F^{ns}_{r,k}$.*

Proof.

$$\begin{aligned} F^{sn}_{r,k} &= \sum_m \frac{1}{(2\pi \mathrm{i})^2} \iint_{C\,C'} A_m(Q)A_r(Q)\omega^s(Q)\omega^n(Q')\omega^m(Q')e_k(Q') \\ &= \frac{1}{(2\pi \mathrm{i})^2} \iint_{C\,C'} A_r(Q)\omega^s(Q)\omega^n(Q')e_k(Q')\Delta(Q, Q') \\ &= \frac{1}{2\pi \mathrm{i}} \oint_C A_r(Q)\omega^s(Q)\omega^n(Q)e_k(Q). \end{aligned}$$

The latter expression is obviously symmetric with respect to n and s. □

Further on, $A_0^{(1)} + B_0^{(1)} = \sum_{n,s} \sum_i \big(:u_i(n)[u^i, x](s) + [u_i, x](n)u^i(s):\big)\, F_{r,k}^{sn}$.

Lemma 2.32. $\sum_i \big(:u_i(n)[u^i, x](s) + [u_i, x](n)u^i(s):\big) = 0$.

Proof. We have

$$\begin{aligned}\sum_i u_i(n)[u^i, x](s) &= \sum_i u_i(n) \sum_j ([u^i, x], u_j)u^j(s) \\ &= -\sum_{i,j} u_i(n)(u^i, [u_j, x])u^j(s) \\ &= -\sum_j [u_j, x](n)u^j(s).\end{aligned}$$

Hence $A_0^{(1)} + B_0^{(1)} = 0$. □

Let us return to (2.8.16).

Proposition 2.33. *For a nonnegative integer* N *and any sufficiently small* ϵ*, the coefficient at* $x(v)$ *on the right-hand side of* (2.8.16) *is given by*

$$E_\epsilon^{(N)} = \sum_{m,s} \sum_{n>N} \alpha_{ns}^v \alpha_{mr}^s l_k^{nm} \psi(\epsilon n) - \sum_{n,m} \sum_{s>N} \alpha_{sm}^v \alpha_{nr}^s l_k^{nm} \psi(\epsilon n) \tag{2.8.17}$$

for an arbitrary v.

Proof. The difference between a summand in (2.8.16) and the expression (2.8.17) with the same index v is equal to[1]

$$\sum_{m,s} \sum_{n=s+1}^{N} \alpha_{ns}^v \alpha_{mr}^s l_k^{nm} \psi(\epsilon n) - \sum_{n,m} \sum_{s=m+1}^{N} \alpha_{sm}^v \alpha_{nr}^s l_k^{nm} \psi(\epsilon n). \tag{2.8.18}$$

By almost-grading, there are only finitely many summands in each of the sums for a fixed triple v, k, r. Hence we may forget about $\psi(\epsilon n)$ and perform the following substitution of variables in the second sum: $(s \to n \to m \to s)$. By Lemma 2.31, we obtain that the difference is equal to zero. □

It follows from the proof that the result does not depend on a normal ordering. Consider

$$E_\epsilon^{(0)} = \sum_s \sum_{n>0} \alpha_{ns}^v F_{r,k}^{sn} \psi(\epsilon n) - \sum_n \sum_{s>0} \alpha_{nr}^s F_{s,k}^{vn} \psi(\epsilon n). \tag{2.8.19}$$

[1] If the upper boundary is less than the lower one, we assume that the signs of the summands change. This convention holds until the end of the section.

Observe that the second sum can be transformed as follows:

$$\sum_{n,s>0} = \sum_{s,n>0} - \sum_{n>0,s\le 0} + \sum_{s>0,n\le 0},$$

hence

$$E_\epsilon^{(0)} = \sum_{n>0}\sum_{s}\left(\alpha_{ns}^{v}F_{r,k}^{sn} - \alpha_{nr}^{s}F_{s,k}^{vn}\right)\psi(\epsilon n) + \left(\sum_{\substack{n>0\\ s\le 0}} - \sum_{\substack{s>0\\ n\le 0}}\right)\alpha_{nr}^{s}F_{s,k}^{vn}\psi(\epsilon n). \tag{2.8.20}$$

By summing up over s in the first double sum and using the delta-distribution, we prove that this double sum is equal to zero. Making use of the integral representation for $F_{s,k}^{vn}$ (Lemma 2.31), and for α_{nr}^{s} (p. 44), the remainder of the expression (2.8.20) can be represented as

$$\frac{1}{(2\pi\,\mathrm{i})^2}\iint\limits_{C\,C'} A_r(Q')e_k(Q)\omega^v(Q)\Bigg(\sum_{\substack{n>0\\ s\le 0}} A_n(Q')A_s(Q)\omega^s(Q')\omega^n(Q)\psi(\epsilon n)$$
$$- \sum_{\substack{s>0\\ n\le 0}} A_n(Q')A_s(Q)\omega^s(Q')\omega^n(Q)\psi(\epsilon n)\Bigg). \tag{2.8.21}$$

Later on we shall make use of the following lemma.

Lemma 2.34 ([4]). *The following relation holds for an arbitrary N:*

$$\left(\sum_{n>N}\sum_{s\le N} - \sum_{s>N}\sum_{n\le N}\right)A_n(Q')A_s(Q)\omega^s(Q')\omega^n(Q) = d'\Delta(Q',Q), \tag{2.8.22}$$

where d' denotes differential in Q'.

Applying the lemma gives

$$\begin{aligned} E_0^{(0)} &= \frac{1}{(2\pi\,\mathrm{i})^2}\iint\limits_{C\,C'} A_r(Q')e_k(Q)\omega^v(Q)d'\Delta(Q',Q) \\ &= -\frac{1}{(2\pi\,\mathrm{i})^2}\iint\limits_{C\,C'} d'A_r(Q')e_k(Q)\omega^v(Q)\Delta(Q',Q) \\ &= -\frac{1}{2\pi\,\mathrm{i}}\oint_C dA_r(Q)e_k(Q)\omega^v(Q) = -K_{r,k}^{v}. \end{aligned}$$

This completes computation of the coefficient at $x(v)$ (for an arbitrary v) in (2.8.16). The long expression for that coefficient on the right-hand side of (2.8.16) may now be replaced with $-K^v_{r,k}$, giving

$$A^{(2)}_\epsilon + B^{(2)}_\epsilon = -2\varkappa \sum_v K^v_{r,k} x(v). \tag{2.8.23}$$

Substituting (2.8.23), together with (2.8.15), to (2.8.13) gives (2.8.1).

To complete the proof of Proposition 2.25, we still have to prove Lemma 2.34. First, let us prove the following relation.

Lemma 2.35.

$$\gamma_{rk} = \left(\sum_{n>0}\sum_{s\le 0} - \sum_{s>0}\sum_{n\le 0}\right)\alpha^s_{rn}\alpha^n_{ks}. \tag{2.8.24}$$

Proof. In [4], it is proposed to make use of semiinfinite forms to prove this lemma. Here we take semiinfinite forms on $\mathcal{A}$. Let us take a vacuum vector $\Phi = A_1 \wedge A_2 \wedge \cdots$. An element $A_i \in \mathcal{A}$ acts by the Leibniz rule, i.e., as follows:

$$A_i\Phi = (A_i \cdot A_1) \wedge A_2 \wedge \cdots + A_1 \wedge (A_i \cdot A_2) \wedge A_3 \cdots + \cdots \tag{2.8.25}$$

If $|i|$ is sufficiently large, it makes sense. For some critical range of indices (in particular, for $A_0 = 1$), the action must be regularized, and we will obtain a representation of a central extension $\widehat{\mathcal{A}}'$ with a local cocycle (Section 1.5). As $\mathcal{A}$ is a commutative Lie algebra, two different cocycles are never cohomologous. As shown on page 33, the cocycle of the fermion representation of the Lie algebra $\mathcal{A}$ is always $\mathcal{V}$-invariant, and hence by Theorem 1.8 it is proportional to the standard cocycle $\gamma(A,B) = \operatorname{res}_{P_+} AdB$. Hence $[A_r, A_k]\Phi = d\cdot\gamma(A_r, A_k)\Phi = d\cdot\gamma_{rk}\Phi$ where $d \in \mathbb{C}$. If r and k are outside the critical range, the action of A_r (respectively, A_k) is given by the relation (2.8.25). Inside the critical range, it is necessary to consider how an element A_s contained in Φ reproduces itself. First, $A_k \cdot A_s = \sum_n \alpha^n_{ks} A_n$. This combination may occur only if $s \ge 1$; in this case only the terms with $n < 1$ will not annihilate with the surrounding elements. Furthere on, acting by A_r, and considering the projection onto the initial element A_s, we obtain $\sum_{s>0}\sum_{n\le 0}\alpha^s_{rn}\alpha^n_{ks}$. Considering $-A_k \cdot A_r$ and replacing the indices, we obtain

$$[A_r, A_k]\Phi = -\left(\sum_{n>0}\sum_{s\le 0} - \sum_{s>0}\sum_{n\le 0}\right)\alpha^s_{rn}\alpha^n_{ks}\Phi = d\cdot\gamma_{rk}\Phi\,. \tag{2.8.26}$$

To find out the constant d, compute this expression for $r = i$ and $k = -i$ where $i \gg 0$. Observe that $A_i\Phi = 0$, hence $[A_i, A_{-i}]\Phi = A_i(A_{-i}\Phi)$. In the two-point case considered here, the general multiplication relation of Proposition 1.2 specializes as $A_k \cdot A_s = A_{k+s} + \sum_{j>k+s} C_j A_j$. Applying this for $k = -i$ we obtain

$$A_{-i}\Phi = \sum_{s-i\le 0} A_1 \wedge \cdots A_{s-1} \wedge A_{s-i} \wedge A_{s+1} \cdots\,.$$

The terms with $s - i > 0$ vanish due to presence of repeated wedge factors.

The consequent action by A_i turns every summand back to Φ. Thus we obtain $A_i A_{-i}\Phi = i\,\Phi$, and further on

$$[A_i, A_{-i}]\Phi = i\,\Phi\,.$$

On the other hand side, we find that $\gamma(A_i, A_{-i}) = -i$ by computing the residues. Hence by (2.8.26) we find that $d = -1$, which completes the proof. □

Remark 2.36. The proof in [4] is based on a straightforward computation.

Proof of Lemma 2.34. We have

$$dA_k(Q) = \sum_r \beta_{kr}\omega^r(Q), \qquad \beta_{kr} = \frac{1}{2\pi\,\mathrm{i}}\oint_C dA_k(Q)A_r(Q) = \gamma_{rk}, \quad (2.8.27)$$

hence by Lemma 2.35

$$d'\Delta(Q', Q) = \sum_k d'A_k(Q')\omega^k(Q) = \sum_{k,r}\Bigg(\sum_{\substack{n>0\\ s\le 0}} - \sum_{\substack{s>0\\ n\le 0}}\Bigg)\alpha^s_{rn}\alpha^n_{ks}\omega^k(Q)\omega^r(Q'). \quad (2.8.28)$$

Making use of the relation $A_i\omega^j = \sum_r \alpha^j_{ir}\omega^r$ we obtained the desired result for $N = 0$. To obtain it for an arbitrary N, it is sufficient to compare the summation range with the one for $N = 0$. The difference is equal to

$$\Bigg(\sum_{s=1}^{N}\sum_{n} - \sum_{n=1}^{N}\sum_{s}\Bigg)\alpha^s_{rn}\alpha^n_{ks}\omega^k(Q)\omega^r(Q').$$

Each of the double sums has finitely many summands, and therefore we may consider them separately. Doing so for the summation over n in the first double sum, and over s in the second one, we obtain (using the integral representation of the coefficients, and the delta-distribution) the same quantities which contract. □

Now the proof of Proposition 2.25 is really completed. Let us prove Proposition 2.27. Again, for $\epsilon \neq 0$ we write

$$\big[L_k(\epsilon), L_l\big] = \frac{1}{2}\sum_{n,m}\sum_i l_k^{nm}\big[\,{:}u_i(n)u^i(m){:}\,,\, L_l\big]\psi(\epsilon n). \quad (2.8.29)$$

As explained at the beginning of the present section, we can ignore the normal ordering inside commutators and write (2.8.29) as

$$\frac{1}{2}\sum_{n,m}\sum_i l_k^{nm}\Big(u_i(n)\big[u^i(m), L_l\big] + \big[u_i(n), L_l\big]u^i(m)\Big)\psi(\epsilon n). \quad (2.8.30)$$

Transform commutators using Proposition 2.25 and obtain

$$\frac{1}{2}(c+\varkappa)\sum_{n,m,v}\sum_i l_k^{nm}\big(K^v_{m,l}u_i(n)u^i(v) + K^v_{n,l}u_i(v)u^i(m)\big)\psi(\epsilon n). \quad (2.8.31)$$

To deal with expressions well-defined for $\epsilon = 0$, let us return to the normal ordering, introducing commutators for those terms which are not in the normal order. Having transformed those commutators following Lemma 2.30 (3), we obtain the sum $A_\epsilon + B_\epsilon + C_\epsilon + D_\epsilon$ as the result where the summands are

$$\begin{aligned}
A_\epsilon &= \frac{1}{2}(c+\varkappa)\sum_{n,m,v}\sum_i l_k^{nm} K_{m,l}^v :u_i(n)u^i(v): \psi(\epsilon n),\\
B_\epsilon &= \frac{1}{2}(c+\varkappa)\sum_{n,m,v}\sum_i l_k^{nm} K_{n,l}^v :u_i(v)u^i(m): \psi(\epsilon n),\\
C_\epsilon &= -\frac{1}{2}c(c+\varkappa)\dim\mathfrak{g}\sum_{v,m}\sum_{n>v} l_k^{nm} K_{m,l}^v \gamma_{nv}\psi(\epsilon n),\\
D_\epsilon &= -\frac{1}{2}c(c+\varkappa)\dim\mathfrak{g}\sum_{n,m}\sum_{v>m} l_k^{nm} K_{n,l}^v \gamma_{vm}\psi(\epsilon n).
\end{aligned} \tag{2.8.32}$$

Considering the range of indices where the coefficients l_k^{nm} and $K_{m,l}^v$ can be nonzero, and taking account of the normal ordering, we observe that A_0 and B_0, again, are well-defined operators, which enables us to ignore the factor $\psi(\epsilon n)$. Renaming indices in B_0 as $(v \to n \to m \to v)$, we obtain

$$A_0 + B_0 = \frac{1}{2}(c+\varkappa)\sum_{n,m,v}\sum_i \left(l_k^{nm} K_{m,l}^v + l_k^{mv} K_{m,l}^n\right) :u_i(n)u^i(v):. \tag{2.8.33}$$

Lemma 2.37.

$$\sum_m \left(l_k^{nm} K_{m,l}^v + l_k^{mv} K_{m,l}^n\right) = -\sum_s C_{kl}^s l_s^{nv}. \tag{2.8.34}$$

where

$$C_{kl}^s = \frac{1}{2\pi\,\mathrm{i}}\oint_C ([e_k, e_l]) \cdot \Omega^s. \tag{2.8.35}$$

are structure constants of the vector field algebra $\mathcal{V}$.

Proof. The right-hand side of (2.8.34) is written as

$$-\sum_s \frac{1}{(2\pi\,\mathrm{i})^2}\iint_{C\,C'} [e_k, e_l](Q)\cdot\Omega^s(Q)\omega^n(Q')\omega^v(Q')e_s(Q').$$

The summation over s results in the delta-distribution of tensor weight $(-1,2)$. By the consequent integration over Q' we obtain

$$-\sum_s C_{kl}^s l_s^{nv} = -\frac{1}{2\pi\,\mathrm{i}}\oint_C [e_k, e_l](Q)\omega^n(Q)\omega^v(Q). \tag{2.8.36}$$

For the first sum on the left hand side of (2.8.34) we have

$$\sum_m l_k^{nm} K_{m,l}^v = \sum_m \frac{1}{(2\pi\,\mathrm{i})^2} \iint_{C\,C'} \omega^n(Q)\omega^m(Q)e_k(Q)d'A_m(Q')e_l(Q')\omega^v(Q').$$

Having applied the relation $\sum_m d'A_m(Q')\omega^m(Q) = d'\Delta(Q',Q)$ and integrated over Q', we obtain

$$\sum_m l_k^{nm} K_{m,l}^v = -\frac{1}{2\pi\,\mathrm{i}} \oint_C \omega^n(Q)e_k(Q)d\big(e_l(Q)\omega^v(Q)\big).$$

For the second sum we have

$$\begin{aligned}\sum_m l_k^{mv} K_{m,l}^n &= -\frac{1}{2\pi\,\mathrm{i}} \oint_C \omega^v(Q)e_k(Q)d\big(e_l(Q)\omega^n(Q)\big)\\ &= \frac{1}{2\pi\,\mathrm{i}} \oint_C d\big(\omega^v(Q)e_k(Q)\big)e_l(Q)\omega^n(Q),\end{aligned}$$

and altogether

$$\begin{aligned}&\sum_m (l_k^{nm} K_{m,l}^v + l_k^{mv} K_{m,l}^n)\\ &= \frac{1}{2\pi\,\mathrm{i}} \oint_C \Big(-\omega^n(Q)e_k(Q)d\big(e_l(Q)\omega^v(Q)\big) + \omega^n(Q)e_l(Q)d\big(e_k(Q)\omega^v(Q)\big)\Big).\end{aligned}$$

In local coordinates we have for the integrand of the last expression

$$-\omega^n(z)\omega^v(z)\big(e_k(z)\frac{de_l}{dz}(z) + e_l(z)\frac{de_k}{dz}(z)\big) = -\omega^n(z)\omega^v(z)[e_k,e_l](z)$$

which, up to the sign, coincides with the integrand in (2.8.36). This completes the proof. □

By (2.8.33) and Lemma 2.37 we obtain

$$\begin{aligned}A_0 + B_0 &= -\frac{1}{2}(c+\varkappa)\sum_s \sum_{n,v} \sum_i C_{kl}^s l_s^{nv} :u_i(n)u^i(v):\\ &= -(c+\varkappa)\sum_s C_{kl}^s L_s.\end{aligned}$$

We still have to examine $\alpha(k,l) = \lim_{\epsilon\to 0}(C_\epsilon + D_\epsilon)$. Since L_k and L_l are well-defined operators in the space V, the scalar $\alpha(k,l)$ is also well-defined. What we have just proven implies that

$$[L_k, L_l] = -(c+\varkappa)\sum_s C_{kl}^s L_s + \alpha(k,l)\cdot id\ .$$

Hence α is a cocycle on $\mathcal{V}$. By examining orders and residues of the 1-forms involved to the definitions of the coefficients l_k^{nm}, $K_{m,l}^v$, and γ_{nv} we obtain that

$$\begin{array}{lll} \text{if } K_{m,l}^v \neq 0 & \text{then} & -3g \leq -v + l + m \leq 0, \\ \text{if } l_k^{nm} \neq 0 & \text{then} & -g \leq k - (n+m) \leq 0, \\ \text{if } \gamma_{nv} \neq 0 & \text{then} & -2g \leq n + v \leq 0. \end{array}$$

Taking this into account, we obtain that if $\alpha(k,l) \neq 0$, then $-6g \leq k + l \leq 0$, i.e., α is a local cocycle. For $g \neq 0$, $n = -g$ and $v = -g - 1$ (or vice verse), $\gamma_{-g,-g-1}$ does not vanish but the result remains true due to other coefficients. The obtained cocycle can be given by

$$\alpha(k,l) = -\frac{1}{2}c(c+\varkappa)\dim\mathfrak{g}\sum_n\Big(\sum_{m,s}\sum_{v<n} l_k^{nm} l_l^{vs}\gamma_{sm}\gamma_{nv} + \sum_{m,s}\sum_{v>m} l_k^{nm} l_l^{vs}\gamma_{sn}\gamma_{vm}\Big).$$

Thus the main statement (2.8.4) of Proposition 2.27 has been proven.

In the remainder of the section we prove the relations (2.8.5)-(2.8.7). Consider again the quantities (2.8.32), for the moment without any normalizing factor $(-1/2)c$ $(c+\varkappa)\dim\mathfrak{g}$. Given an integer N, consider first C_ϵ and D_ϵ. Set

$$E_\epsilon = \sum_{v,m}\sum_{n>N} l_k^{nm} K_{m,l}^v \gamma_{nv}\psi(\epsilon n), \quad F_\epsilon := \sum_{n,m}\sum_{v>N} l_k^{nm} K_{n,l}^v \gamma_{vm}\psi(\epsilon n).$$

There are only finitely many summands in $C_\epsilon - E_\epsilon$, and $D_\epsilon - F_\epsilon$. Hence we may set $\epsilon = 0$, and forget of $\psi(\epsilon n)$. Let us rename the summation indices in $D_\epsilon - F_\epsilon$ as follows: $v \to n \to m \to v$, then by Lemma 2.37 we will obtain the following expression:

$$\begin{aligned} \lim_{\epsilon\to 0}\big((C_\epsilon - E_\epsilon) + (D_\epsilon - F_\epsilon)\big) &= \sum_v \sum_{n=v+1}^{N}\Big(\sum_m l_k^{nm} K_{m,l}^v + l_k^{mv} K_{ml}^n\Big)\gamma_{nv} \\ &= -\sum_v \sum_{n=v+1}^{N}\sum_s C_{kl}^s l_s^{nv}\gamma_{nv}. \end{aligned} \tag{2.8.37}$$

Unlike in the case which took place in the proof of Proposition 2.25, this expression does not vanish. I particular, it is one of the reasons why the cocycle depends on a normal ordering.

Further on, let us consider $E_\epsilon + F_\epsilon$. Present the summation range in F_ϵ as $(n, v > N) = (v, n > N) + (v > N, n \leq N) - (n > N, v \leq N)$, and denote the first sum by $F_\epsilon^{(1)}$, and two others together by $F_\epsilon^{(2)}$. First, using the relations $\gamma_{vm} = -\gamma_{mv}$, and $K_{v,k}^n = \sum_m l_k^{nm}\gamma_{mv}$ we obtain

$$E_\epsilon + F_\epsilon^{(1)} = \sum_v\sum_{n>N}\Big(\Big(\sum_m l_k^{nm} K_{m,l}^v\Big)\gamma_{nv} - K_{n,l}^v K_{v,k}^n\Big)\psi(\epsilon n).$$

Since $F_\epsilon^{(2)}$ is well defined for $\epsilon = 0$ (see below), we may ignore $\psi(\epsilon n)$ here if we do not split it. Making use of the integral representation of the coefficients, and summing up over m, we obtain the delta-distribution (as was already done several times), and further on

$$\begin{aligned}\Big(\sum_m l_k^{nm} K_{m,l}^v\Big)\gamma_{nv} \\ &= -\frac{1}{(2\pi\,\mathrm{i})^2}\iint\limits_{C\,C'} w^v(Q)d'A_n(Q')e_l(Q)d\big(e_k(Q)\omega^n(Q)\big)A_v(Q'), \\ K_{n,l}^v K_{v,k}^n &= \frac{1}{(2\pi\,\mathrm{i})^2}\iint\limits_{C\,C'} w^v(Q')d'A_n(Q')e_l(Q')d\big(e_k(Q)\omega^n(Q)\big)A_v(Q).\end{aligned}$$

By summation over v, and integration over Q' we obtain

$$E_\epsilon + F_\epsilon^{(1)} = 0.$$

Consider now

$$F_\epsilon^{(2)} = \Big(\sum_{n>N}\sum_{v\le N} - \sum_{n\le N}\sum_{v>N}\Big)K_{v,k}^n K_{n,l}^v\,\psi(\epsilon n). \tag{2.8.38}$$

Each of the double sums consists of only finitely many summands; hence we may again set $\epsilon = 0$. For $N = 0$ we gather the corresponding terms of (2.8.37) and (2.8.38), and obtain the desired result (2.8.5) on the structure of the cocycle, having been formulated in Proposition 2.27. Having chosen another normal ordering, we would obtain a different summation rule in (2.8.37). The expression (2.8.38) would be invariant. One more thing that must be done is to derive the relation (2.8.7) for the cocycle values on special basis elements. First, consider $\widehat{\chi}_{k,-k}$. Here $K_{v,k}^n \neq 0$ implies that $k \le n - v$, and $K_{n,-k}^v \neq 0$ implies $-k \le v - n$, hence $v = n - k$. Thus, we have obtained $(n-k)$ in this case (respectively n). Assume, $k \ge 0$, then only the first sum will survive. It is finite and is written as

$$\sum_{n=1}^{k}(n-k)n = -\frac{1}{6}(k^3 - k)\,.$$

In the expression for $\psi_{k,-k}$ we find that $C_{k,-k}^s l_s^{nv}\gamma_{nv} \neq 0$ only if $s \ge 0$, $s \le n + v$, and $n + v \le 0$. This means that we have nonvanishing terms only for $s = 0$ and $n = -v$. But examining the summation $\sum_v \sum_{n=0}^{v+1}$ we observe that $n = -v$ implies that either $v = 0$ (with a zero coefficient), or the sumation range is empty. Hence $\psi_{k,-k} = 0$. Altogether, we obtain (2.8.7).

Let us conclude with the following instructive statement.

Corollary 2.38. *Given an admissible representation of* $\overline{\mathfrak{g}}$, *and an arbitrary basis element* $e_l \in \mathcal{V}$, *let* E *be an operator satisfying the commutation relations*

$$\big[E, x(r)\big] = -(c + \varkappa)\, x(\nabla_{e_l} A_r)\,, \tag{2.8.39}$$

for any $x \in \mathfrak{g}$, *and any* r. *Then for any* k

$$[L_k, E] = [L_k, L_l]\,.$$

Proof. The proof of Proposition 2.27 is based on the relation $[L_l, x(r)] = -(c + \varkappa)x \otimes (\nabla_{e_l} A_r)$ of Proposition 2.26. Nothing changes if L_l is replaced with E. Following the same lines, we obtain $[L_k(\epsilon), E] = [L_k(\epsilon), L_l]$; hence in the limit $\epsilon \to 0$ we will have $[L_k, E] = [L_k, L_l]$. □

Chapter 3

Projective flat connections on the moduli space of punctured Riemann surfaces and the Knizhnik–Zamolodchikov equation

This chapter deals with the applications of the Krichever–Novikov algebras and their representations to the geometry of moduli spaces of Riemann surfaces, and 2-dimensional conformal field theory (2D CFT).

These applications are based on Proposition 3.3 and its corollary, together providing a description of tangent spaces to moduli spaces of Riemann surfaces in terms of Krichever–Novikov vector field algebras and their bases. The tangent space to the moduli space is related to to the vector field algebra by means the Kodaira–Spencer mapping, and its inverse.

Starting from [8], the 2D CFT is identified with a projective flat connection on a moduli space of Riemann surfaces, given by the energy-momentum tensor. In the particular 2D CFT model referred to as the Wess–Zumino–Novikov–Witten model (WZNW), the connection can be given by a certain expression, which is called the *Knizhnik–Zamolodchikov connection* in this case. The Wess–Zumino–Novikov–Witten model is distinguished among the other 2D CFT models by the specific relation between the energy-momentum tensor and the current. This relation is nothing but the Sugawara construction. We give here a self-contained mathematical treatment of the Knizhnik–Zamolodchikov connection which includes the following steps. First, we construct the sheaf of *conformal blocks* on the moduli space (more precisely, the sheaf of *coinvariants*) by means the representation theory of affine Krichever–Novikov algebras. Then we give a general expression for the Knizhnik–Zamolodchikov connection in terms of fundamental objects, namely the energy-momentum tensor and the Kodaira–Spencer mapping (3.4.1). In itself, this expression makes sense in the general 2D CFT setup, but we prove it only for WZNW. Two problems have to be resolved on the way: well-definiteness of the connection on coinvariants, and its projective flatness. The solutions to both problems are based on the important technical tool: the deformation formulae of regular Krichever–Novikov functions and vector fields (Section 3.3).

To conclude, we explicitly derive the Knizhnik–Zamolodchikov equations for genus 0 and genus 1 and show that in the first case they coinside with the conventional Knizhnik–Zamolodchikov equations.

The Knizhnik–Zamolodchikov equations were obtained in [19], initially for genus 0. The first generalization to positive genera was undertaken by D. Bernard in [2,3]. In those works D. Bernard introduced additional parameters (twists) and conventions not assumed by foundations of CFT. The equations he obtained are referred to as Knizhnik–Zamolodchikov–Bernard equations. Many works, including those by D. Ivanov, G. Felder, and others follow this direction. We follow another direction given in [74] by A. Tsuchia, K. Ueno, and Y. Yamada (see also [77]). Unlike them, we work with Krichever–Novikov algebras as algebras of gauge and conformal symmetries of the theory. It can be observed from the original work [19] that the WZNW action functional is indeed invariant with respect to those Lie algebras.

A counterpart of the Knizhnik–Zamolodchikov equations is given by N. Hitchin in [12]. The Hitchin connection contains a potential of the Ricci curvature. It is a differential geometric, and not a conformal invariant object. It is obtained for the moduli spaces of curves without marked points.

This chapter is mainly based on the papers [52, 53] (see also [65]). There are a number of monographs treating those faces of CFT which are not considered here, for example [55,76].

3.1 Virasoro-type algebras and moduli spaces of Riemann surfaces

In this section, following [52,53], we describe the moduli spaces of Riemann surfaces needed for 2D CFT.

Let $\mathcal{M}^{(k,p)}_{g,N+1}$ denote the moduli space of Riemann surfaces of genus g with $N+1$ different ordered marked points, fixed k-jets of local coordinates at the first N points, and fixed p-jet of a local coordinate at the last point. The elements of the space $\mathcal{M}^{(k,p)}_{g,N+1}$ are given by sets of the form

$$\tilde{b}^{(k,p)} = \left[\Sigma, P_1, \ldots, P_N, P_\infty, z_1^{(k)} \ldots, z_N^{(k)}, z_\infty^{(p)}\right], \tag{3.1.1}$$

where Σ is a smooth projective curve of genus g, P_i, and ($i = 1, \ldots, N, \infty$) are different points at Σ, z_i is a local coordinate at the point P_i (such that $z_i(P_i) = 0$), and $z_i^{(l)}$ is an l-jet of the local coordinate z_i ($l \in \mathbb{N} \cup \{0\}$). Here, [..] denotes a class of such sets equivalent in the following sense. Two sets $\tilde{b}^{(k,p)}$ and $\tilde{b}^{(k,p)\prime}$ are equivalent if there exists an algebraic isomorphism $\phi : \Sigma \to \Sigma'$ with $\phi(P_i) = P_i'$ for $i = 1, \ldots, N, \infty$, such that after identification by means ϕ we obtain

$$z_i' = z_i + O(z_i^{k+1}), \quad i = 1, \ldots, N \quad \text{and} \quad z_\infty' = z_\infty + O(z_\infty^{p+1}). \tag{3.1.2}$$

Let us introduce a special notation in the following cases: $\mathcal{M}_{g,N+1} = \mathcal{M}^{(0,0)}_{g,N+1}$, and $\mathcal{M}^{(1)}_{g,N+1} = \mathcal{M}^{(1,0)}_{g,N+1}$. "Forgetting" coordinates, or higher-order jets, we obtain the following natural projections

$$\mathcal{M}_{g,N+1}^{(1,p)} \to \mathcal{M}_{g,N+1}, \qquad \mathcal{M}_{g,N+1}^{(k,p)} \to \mathcal{M}_{g,N+1}^{(k',p')} \tag{3.1.3}$$

for any $k' \leq k$, and $p' \leq p$.

In the neighborhood of a generic point (i.e., the point corresponding to generic curve and marked points), $\mathcal{M}_{g,N+1}^{(k,p)}$ is smooth. Let us give a description of the tangent space to $\mathcal{M}_{g,N+1}^{(k,p)}$ at such point. For an arbitrary divisor $D = \sum_{P \in \Sigma} n_P P$ ($n_P \in \mathbb{Z}$, and, in addition, $n_P \neq 0$ for only finitely many points), let $T_\Sigma(D)$ denote the sheaf of (local) meromorphic vector fields e, satisfying the constraint $\operatorname{ord}_P e \geq -n_P$ for all $P \in \Sigma$ (which means that e is holomorphic outside the supp D, in particular). By the Kodaira–Spencer theory (see, e.g., [20]) the tangent space may be identified with the cohomology space with coefficients in this kind of sheaf:

$$\mathrm{T}_{\tilde{b}^{(k,p)}} \mathcal{M}_{g,N+1}^{(k,p)} \cong \mathrm{H}^1\Big(\Sigma, T_\Sigma\big(-(k+1)S - (p+1)P_\infty\big)\Big), \tag{3.1.4}$$

where $S = \sum_{i=1}^N P_i$ is the divisor of the marked points on Σ. The isomorphism (3.1.4) is referred to as the *Kodaira–Spencer isomorphism*, and the cohomology group on the right-hand side of (3.1.4) as the *Kuranishi tangent space*. There are natural mappings of the Krichever–Novikov vector field algebra $\mathcal{V}$ onto both spaces involved into the relation (3.1.4). First, define the following mapping:

$$\Theta : \mathcal{V} \to \mathrm{H}^1\Big(\Sigma, T_\Sigma\big(-(k+1)S - (p+1)P_\infty\big)\Big) \tag{3.1.5}$$

originally introduced in [52].

Let U_∞ be a coordinate neighborhood of the point P_∞ containing none of the points $P_1, \ldots, P_N$. Set $U_0 = \Sigma \setminus \{P_\infty\}$, $U = U_\infty \cap U_0 = U_\infty \setminus \{P_\infty\}$.

For any $e \in \mathcal{V}$ let us set by definition

$$\theta(e) = e|_U, \tag{3.1.6}$$

i.e., θ is just a restriction of the vector field to U. Let us regard to $\theta(e)$ as to the 1-cocycle in the sense of a Chech cohomology. Let

$$\Theta(e) = [\theta(e)] \tag{3.1.7}$$

be the corresponding cohomology class.

Lemma 3.1. *The mapping Θ is surjective. For its kernel we have*

$$\ker \Theta = \mathcal{V}^{(k+1)} \oplus \mathcal{V}^{reg}, \tag{3.1.8}$$

where $\mathcal{V}^{(k+1)} = \mathcal{V}_k \oplus \mathcal{V}_{k+1} \oplus \ldots$ and $\mathcal{V}^{reg}$ is the subspace in $\mathcal{V}$ consisting of the vector fields regular at P_∞ and vanishing at that point with the order at least $p+1$. Both $\mathcal{V}^{(k+1)}$ and $\mathcal{V}^{reg}$ are Lie subalgebras in $\mathcal{V}$.

Proof. Let the vector field e belong to the kernel of the mapping Θ. Then it gives a coboundary and hence may be represented as $e = e_\infty - e_0$ in the domain U where

e_∞ is defined and holomorphic in U_∞, and has the order $p+1$, at least, at P_∞, e_0 is defined and holomorphic in U_0, and has the order $k+1$, at least, at every point $P_1,\dots,P_N$.

Since $e_0 = e - e_\infty$ where e is a global meromorphic vector field, and e_∞ is holomorphic at P_∞, e_0 has the singularity of the same type as e at P_∞. For this reason e_0 is also a global meromorphic vector field. For the same reason e_∞ is also a global meromorphic vector field. By our assumptions on the orders of e_0 and e_∞ at the marked points, we conclude that

$$e_0 \in \mathcal{V}^{(k+1)}, \; e_\infty \in \mathcal{V}^{reg}. \tag{3.1.9}$$

We have shown that $\ker\Theta \subseteq \mathcal{V}^{(k+1)} \oplus \mathcal{V}^{reg}$. By the definition of cohomology, we immediately obtain the inverse inclusion. Thus (3.1.8) is proven.

Let us calculate the dimension of the quotient space $\mathcal{V}/\ker\Theta$. Its dual is canonically isomorphic to the annihilator of the subspace $\ker\Theta$. By the Krichever–Novikov duality (see Section 1.2), the last consists of quadratic differentials having orders $-(k+1)$, at least, at the points $P_1,\dots,P_N$, and $-(p+1)$, at least, at P_∞. In other words, it consists of quadratic differentials Ω, such that $(\Omega) \geq -((k+1)S + (p+1)P_\infty)$. Quadratic differentials are sections of the bundle $\mathcal{K}^2$. The bundle degree is equal to $2(2g-2)$. By the Riemann-Roch theorem we obtain

$$\begin{aligned}\dim &\left\{\Omega \in \Gamma(\mathcal{K}^2) | (\Omega) \geq -\big((k+1)S + (p+1)P_\infty\big)\right\} \\ &= 2(2g-2) + (k+1)N + (p+1) - (g-1) \\ &= 3g - 2 + (k+1)N + p.\end{aligned}$$

The latter is simply the dimension of the space $\mathcal{M}^{(k,p)}_{g,N+1}$ at a generic point. Indeed, a generic point of $\mathcal{M}^{(k,p)}_{g,N+1}$ is given by $3g-3$ moduli parameters of the Riemann surface, $N+1$ points on the latter, k jet coefficients at each of N points, and p coefficients at P_∞, altogether $3g-3+(N+1)+kN+p$ parameters. Thus

$$\dim(\mathcal{V}/\ker\Theta) = \dim \mathrm{T}_{\tilde{b}^{(k,p)}} \mathcal{M}^{(k,p)}_{g,N+1}. \tag{3.1.10}$$

By (3.1.4) we obtain

$$\dim(\mathcal{V}/\ker\Theta) = \dim \mathrm{H}^1\big(\Sigma, T_\Sigma(-(k+1)S - (p+1)P_\infty)\big). \tag{3.1.11}$$

This proves that the mapping Θ is surjective. □

Lemma 3.1 is formulated and proved in [53] (see [52, Proposition 4.4] for its version with a "fixed" point P_∞). It goes back to [26], where the relation between the Lie algebra $\mathcal{V}$, and the moduli spaces is pointed out in the case of two marked points, and further on, to [21].

Let us proceed with the definition of the mapping

$$\gamma : \mathcal{V} \to \mathrm{T}_{\tilde{b}^{(k,p)}} \mathcal{M}^{(k,p)}_{g,N+1}, \tag{3.1.12}$$

introduced in [11] for the Virasoro algebra and going back to the works of Schiffer and Spencer.

Given a point $\tilde{b}^{(k,p)} \in \mathcal{M}_{g,N+1}^{(k,p)}$, choose a local coordinate w in the neighborhood of the point P_∞ on the Riemann surface Σ corresponding to $\tilde{b}^{(k,p)}$. Assume that w ranges over a disc of radius 2 centered at $w = 0$. Let $U_\infty \subset \mathbb{C}$ be a unit disc with a coordinate z. For $z = w$ (identical transition function) we may think of U_∞ as of a subset of the coordinate chart at the point P_∞. Given an element $e \in \mathcal{V}$, let us associate it with a one-parameter family of transition functions $w = d_t(z)$ $(t \geq 0)$ by means of the relation $d_t(z) = (\exp te)(z)$. Here, $\exp te$ is a local flow corresponding to the vector field e. The transition function d_t is defined in an annulus U_t centered at $z = 0$. In this way, we obtain a one-parameter family of conformal structures, i.e., a curve in the space $\mathcal{M}_{g,N+1}^{(k,p)}$. Introduce local coordinates (*moduli, modular parameters*) $\tau = (\tau_1, \dots, \tau_m)$ in the neighborhood of $\tilde{b}^{(k,p)} \in \mathcal{M}_{g,N+1}^{(k,p)}$, where $m = 3g - 2 + (k+1)N + p$ (the coordinates of the point $\tilde{b}^{(k,p)}$ itself denote by τ_0). With their help, write the constructed curve in $\mathcal{M}_{g,N+1}^{(k,p)}$ in the form $\tau = \tau^e(t)$.

Definition 3.2. Let γ associate a vector field e with the tangent vector to the curve $\tau^e(t)$ at $t = 0$, i.e.,

$$\gamma(e) = \sum_{i=1}^{m} X_i \frac{\partial}{\partial \tau_i}\Big|_{\tau_0}, \quad \text{where } X_i = \frac{d\tau_i^e(t)}{dt}\Big|_{t=0}. \tag{3.1.13}$$

In the end of the section we will show that every tangent vector to the moduli space $\mathcal{M}_{g,N+1}^{(k,p)}$, at generic points, can be obtained in this way. It implies, in particular, that by means of that procedure we obtain all conformal structures which are close to a given one. For this purpose we need a more detailed information of the Kodaira–Spencer isomorphism (3.1.4).

Let us denote the Kodaira–Spencer isomorphism by ρ:

$$\rho : \mathrm{T}_{\tilde{b}^{(k,p)}} \mathcal{M}_{g,N+1}^{(k,p)} \to \mathrm{H}^1\Big(\Sigma, T_\Sigma\big(-(k+1)S - (p+1)P_\infty\big)\Big). \tag{3.1.14}$$

As earlier, let τ_0 be a set of local coordinates of the point $\tilde{b}^{(k,p)}$. Let us consider a family of transition functions $w = d_\tau(z)$, where τ ranges a disc centered at the point τ_0 in the moduli space, and z ranges an annulus in U_∞ centered at 0. The width of this annulus depends on τ.

In general, to define the mapping ρ one needs an advanced cohomological technique. However, at the points where the moduli space is smooth, ρ can be defined purely analytically; let $X \in \mathrm{T}_{\tilde{b}^{(k,p)}} \mathcal{M}_{g,N+1}^{(k,p)}$ be a tangent vector, and $\tau_X(t)$ be a curve in $\mathcal{M}_{g,N+1}^{(k,p)}$ leaving the point $\tilde{b}^{(k,p)}$ in the direction of X. Choose the corresponding transition functions $d_t = d_{\tau_X(t)}$. Set by definition

$$\rho(X) = d_\tau^{-1} \cdot \partial_X d_\tau. \tag{3.1.15}$$

The arrangement of the objects just defined is illustrated by the following figure:

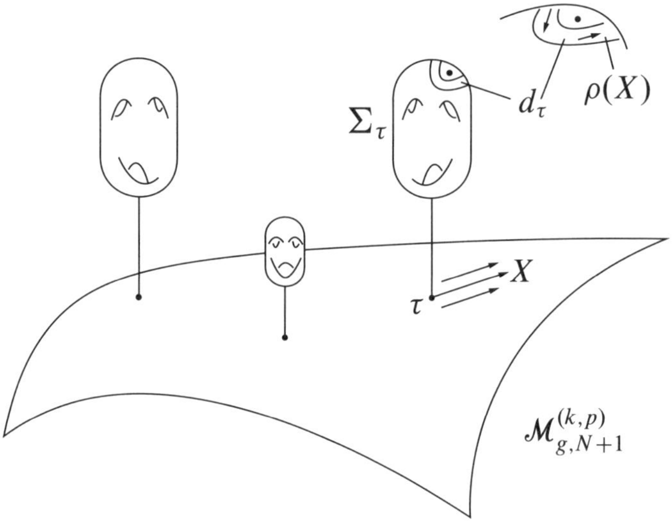

Figure 3.1. Tautological bundle, and the Kodaira–Spencer cocycle.

We regard $\rho(X)$ as a local vector field on the Riemann surface Σ (corresponding to the point $\tilde{b}^{(k,p)}$ of the moduli space, i.e., modular parameters τ_0). It can be explained in two ways. The first explanation is as follows. Every transition function d_τ may be regarded to as a local analytic diffeomorphism with an annulus domain of definition. Let us identify every two locally coinciding diffeomorphisms. Then such diffeomorphisms form a group. In a conventional way, $\rho(X)$ is a tangent vector at the unit of that group, i.e., a local vector field. The relation (3.1.15) provides an invariant definition of that vector field. Let us now give another explanation by means of local coordinates, at the same time making more precise what we mean by the expression $\partial_X d_\tau$. Introduce the following notation: $d(z,\tau) = d_\tau(z)$, and set

$$\partial_X d_\tau(z) = \sum_i X_i \frac{\partial d(z,\tau)}{\partial \tau_i}\Big|_{\tau_0}. \tag{3.1.16}$$

Then $\partial_X d_\tau$ is a local mapping of the 1-dimensional z-space to the w-space. It gives a tangent vector to the group of local diffeomorphisms at the point d_τ. This tangent vector is obtained as follows. We have embedded the family of local diffeomorphisms d_τ into the space of local functions, and taken a tangent vector to that family in the target space.

Then, set

$$\rho_z(X) = d_\tau^{-1}\big(\partial_X d_\tau(z)\big), \quad \text{and} \quad \rho(X) = \rho_z(X)\frac{\partial}{\partial z}. \tag{3.1.17}$$

Here, $\rho_z(X)$ is a local function, and $\rho(X)$ is a local vector field in the z-coordinate representation.

A local vector field defined in the annulus around the point P_∞ gives a Cech 1-cohomology class of the Riemann surface Σ with coefficients in the tangent sheaf. This cocycle is referred to as the *Kodaira–Spencer cocycle*, and its cohomology class is referred to as the *Kodaira–Spencer class* of the 1 parameter family τ_X of deformations of the complex structure on the Riemann surface with marked points. In this form the Kodaira–Spencer class is used e.g. in [77, Lemma 1.3.8]. By misuse of the notation, let $\rho(X)$ stand for the Kodaira–Spencer class.

Proposition 3.3. *The following commutative diagram takes place:*

$$\begin{array}{ccc} \mathcal{V} & \xrightarrow{\ \Theta\ } & \mathrm{H}^1\big(\Sigma, T_\Sigma(-(k+1)S-(p+1)P_\infty)\big) \\ & \searrow^{\gamma} & \uparrow{\scriptstyle\rho} \\ & & \mathrm{T}_{\tilde{b}(k,p)}\mathcal{M}^{(k,p)}_{g,N+1} \end{array} \tag{3.1.18}$$

Proof. Set $e \in \mathcal{V}$, $X = \gamma(e)$. Then, according to (3.1.13), X is defined as a tangent vector at the initial point of the curve $\tau^e(t)$ which corresponds to the one-parameter family of transition functions $d_t(z) = (\exp te)(z)$. Since $\partial_X d_\tau(z)$ in (3.1.16) can be calculated as a derivative in t along the curve, we have $\partial_X d_\tau(z) = \frac{d}{dt}(\exp te)\big|_{t=0} = e$ where e is considered in the local z-coordinate representation. Since $d_0(z)$ is the identity substitution, we have $\rho(X) = e|_U$. By (3.1.6), $\rho(X) = \theta(e)$. Passing to the Cech cohomology classes given by the both parts of the last equality, we have, by virtue of (3.1.7), $\rho(X) = \Theta(e)$. This completes the proof of the commutativity of the diagram (3.1.18). □

Corollary 3.4. *The mapping γ is surjective.*

This assertion follows from Lemma 3.1, and from the fact that ρ is an isomorphism.

Corollary 3.5. *Let $\mathcal{V}_{(0)}$ be an arbitrary complement to* $\ker\Theta$ *in $\mathcal{V}$. Then*

$$\mathrm{T}_{\tilde{b}(k,p)}\mathcal{M}^{(k,p)}_{g,N+1} \cong \mathcal{V}_{(0)}. \tag{3.1.19}$$

Independently of the choice of the complement $\mathcal{V}_{(0)}$, for any $X \in \mathrm{T}_{\tilde{b}(k,p)}\mathcal{M}^{(k,p)}_{g,N+1}$ there exists $e \in \mathcal{V}_{(0)}$ such that $\rho(X) = \Theta(e)$.

The proof of the relation (3.1.19) is obvious. The second assertion of the corollary immediately follows from the commutative diagram (3.1.18).

Remark 3.6. Below, we always assume that the cocycle $\rho(X)$ is represented by a (global) Krichever–Novikov vector field, as established by the last corollary, and write it as:

$$\rho(X) \in \mathcal{V}.$$

In concrete situations, the spaces $\mathcal{V}_{(0)}$ are easily described in terms of the Krichever–Novikov base in vector fields. For example, in the cases $g = 0$ and $g = 1$ this is done in Section 3.5, in the course of constructing the explicit form of the Knizhnik–Zamolodchikov equations.

3.2 Sheaf of conformal blocks and other sheaves on the moduli space $\mathcal{M}^{(1,0)}_{g,N+1}$

Let us consider the moduli space $\mathcal{M}^{(1,0)}_{g,N+1}$. Let $\tilde{b} \in \mathcal{M}^{(1,0)}_{g,N+1}$ be a generic point, and $\widetilde{W}$ be its neighborhood. In analogy with (3.1.1) let us write $\tilde{b}$ as

$$\tilde{b} = \left[\Sigma, P_1, \dots, P_N, P_\infty, z_1^{(1)} \dots, z_N^{(1)}\right]. \tag{3.2.1}$$

Recall that the marked points and 1-jets of local coordinates are exactly the data which make it possible to construct Krichever–Novikov algebras and their fermion representations (Sections 1.1, 2.4, and 2.6).

Let us denote the Krichever–Novikov algebras of various types corresponding to the point $\tilde{b}$ by

$$\mathcal{A}_{\tilde{b}},\ \mathcal{V}_{\tilde{b}},\ \widehat{\mathcal{V}}_{\tilde{b}},\ \overline{\mathfrak{g}}_{\tilde{b}},\ \widehat{\mathfrak{g}}_{\tilde{b}}. \tag{3.2.2}$$

Denote the sheaves of the corresponding objects over $\widetilde{W}$ by

$$\mathcal{A}_{\widetilde{W}},\ \mathcal{V}_{\widetilde{W}},\ \widehat{\mathcal{V}}_{\widetilde{W}},\ \overline{\mathfrak{g}}_{\widetilde{W}},\ \widehat{\mathfrak{g}}_{\widetilde{W}}. \tag{3.2.3}$$

In Section 3.1 we already introduced the regular subalgebra $\mathcal{V}^{reg}$ in the Lie algebra $\mathcal{V}$. For $p = 0$ (the only case considered in this section), $\mathcal{V}^{reg}$ is the subalgebra of the vector fields vanishing at P_∞. A similar subalgebra in $\mathcal{A}$ consisting of the functions vanishing at P_∞ is also called regular. It is denoted by $\mathcal{A}^{reg}$. By definition, $\mathfrak{g}^{reg} = \mathfrak{g} \otimes \mathcal{A}^{reg}$. For any two elements of the space $\mathfrak{g}^{reg}$ the central component of their commutator is equal to zero, since the integrand in (1.6.2) is regular. Hence $\mathfrak{g}^{reg}$ is a subalgebra both in $\overline{\mathfrak{g}}$ and $\widehat{\mathfrak{g}}$ and is called a *regular subalgebra in* $\widehat{\mathfrak{g}}$. Denote by $\mathfrak{g}^{reg}_{\widetilde{W}}$, $\mathcal{V}^{reg}_{\widetilde{W}}$ the corresponding sheaves of regular subalgebras.

Let V be an arbitrary irreducible almost-graded $\widehat{\mathfrak{g}}$-module generated by a vacuum. By a vacuum we mean here an element in V annihilated by the action of all elements in $\overline{\mathfrak{g}}$ regular (as meromorphic functions) at the points $P_1, \dots, P_N$, except for constant matrices, and by constant matrices belonging to the upper nilpotent subalgebra in $\mathfrak{g}$. We assume that V depends on the moduli parameters in general, as well as the algebra itself. We assume that the modules V form a locally free sheaf over an open subset of the moduli space $\mathcal{M}_{g,N+1}^{(1,0)}$. We denote that sheaf by $V_{\widetilde{W}}$.

Definition 3.7. The sheaf of *conformal blocks* associated with the sheaf $V_{\widetilde{W}}$ is defined as the quotient sheaf

$$C_{\widetilde{W}} = V_{\widetilde{W}} / \mathfrak{g}_{\widetilde{W}}^{reg} V_{\widetilde{W}}. \tag{3.2.4}$$

Elements of fibers of the sheaf $C_{\widetilde{W}}$ are referred to as *coinvariants* of the corresponding regular subalgebras.

In the present section, we take the sheaves of fermion modules and twisted Verma modules as $\mathcal{V}_{\widetilde{W}}$, which, however, is not necessary in general. In the remainder of the section, everything is true in a wider setup. It is sufficient that Lemma 3.15 holds.

In our description, the basis fermions do not depend on moduli; only the action of Lie algebras depends on them via the structure constants. We trivialize the sheaf $V_{\widetilde{W}}$ using those bases and obtain the corresponding trivial vector bundle $V \times \widetilde{W}$, where V is a standard fermion space (Section 2.4). Over a generic point of the moduli space, the space $\mathfrak{g}_{\widetilde{W}}^{reg} V$ does not also depend on the moduli – the dependence on the structure constants is eliminated by taking the linear span of images of the basis fermions under the action of $\mathfrak{g}^{reg}$. Hence the sheaf of conformal blocks is locally free and defines a vector bundle. It could be assumed that $\mathfrak{g}$ is an arbitrary reductive Lie algebra. For the sake of simplicity, we will take $\mathfrak{g} = \mathfrak{gl}(n)$, and its standard representation in the n-dimensional space as τ (p. 24). Moreover, coinvariants are finite-dimensional, i.e., the vector bundle of conformal blocks has a finite rank.

Remark 3.8. A definition of the conformal blocks on the space $\widetilde{W}^{(k,p)}$ of generic points in $\mathcal{M}_{g,N+1}^{(k,p)}$ can be given as follows. There is a natural projection $\mathcal{M}_{g,N+1}^{(k,p)} \to \mathcal{M}_{g,N+1}^{(1,0)}$ which simply means "forgetting" the "extra data on jets", namely the components of the order higher than 1 at $P_1, \dots, P_N$, and the whole jet at P_∞. Let us pull back the sheaf $V_{\widetilde{W}}$ from the space $\widetilde{W}$ to the space $\widetilde{W}^{(k,p)}$ by means of that projection. We denote the thusly obtained sheaf by $V^{(k,p)}$. The regular subalgebra $\mathfrak{g}^{(k,p)} \subset \widehat{\mathfrak{g}}$ depends on k and p now. We do not need its description in its full generality. In analogy with (3.2.4) we set by definition

$$C_{\widetilde{W}^{(k,p)}} = V^{(k,p)} / \mathfrak{g}^{(k,p)} V^{(k,p)}. \tag{3.2.5}$$

3.3 Differentiation of the Krichever–Novikov objects in modular variables

The purpose of this section is to define a differentiation of functions and vector fields, and also their representation operators in modular parameters. Observe that this problem is not trivial, thus permitting various nonequivalent approaches. Our approach is compatible with the Kodaira–Spencer description of the tangent space to the moduli space given in terms of the Lie algebra $\mathcal{V}$ (Proposition 3.3, and its corollaries).

As in the previous Section 6.5, we consider the set $\widetilde{W}$ of generic points in $\mathcal{M}^{(1,0)}_{g,N+1}$. Every such point is represented by a set of geometrical data (see (3.2.1))

$$\tilde{b} = \left[\Sigma, P_1, \ldots, P_N, P_\infty z_1^{(1)} \ldots, z_N^{(1)}\right],$$

such that all its components depend on the set of the modular parameters τ (short notation: $\tilde{b} = \tilde{b}(\tau)$). As done in Section 3.1, we associate τ with a gluing function $w = d_\tau(z)$ giving the conformal structure on the Riemann surface $\Sigma(\tau)$. Let us introduce a necessary notation.

Let us choose a point in $\widetilde{W}$ with the modular parameters τ_0, and give a definition of the differentiation of Krichever– Novikov functions in the neighborhood of that point. Such differentiation results in a section of another sheaf, namely the sheaf of the *local functions* on U.

For $\tau \in \widetilde{W}$ let $\mathcal{A}_\tau$ denote the fiber of the functional sheaf $\mathcal{A}_{\widetilde{W}}$ over τ, and let $\mathcal{V}_\tau$ be the same for $\mathcal{V}_{\widetilde{W}}$. Let $\mathcal{A}^{reg}$, $\mathcal{V}^{reg}$, $\mathfrak{g}^{reg}$ denote the corresponding sheaves of the regular subalgebras.

If A is a section of $\mathcal{A}_{\widetilde{W}}$ then we will write its value at the point τ as $A(\tau) = A_\tau$ where $A_\tau \in \mathcal{A}_\tau$. Thus, A_τ is a Krichever–Novikov function. Let us restrict it to U. The obtained local function can be regarded as a function of the local coordinate w, as well as of the local coordinate z. Let us write the result as $A_\tau(w)$ in the first case, and as $\tilde{A}_\tau(z)$ in the second. Thus, we have $\tilde{A}_\tau(z) = A_\tau(d_\tau(z))$. Let $\tilde{\mathcal{A}}_{\widetilde{W}}$ be a sheaf of functions on U (regarded as functions of the z-coordinate). The correspondence $A_\tau \to \tilde{A}_\tau$ gives an embedding of $\mathcal{A}_{\widetilde{W}}$ in $\tilde{\mathcal{A}}_{\widetilde{W}}$. Denote the subsheaf of sections of the sheaf $\tilde{\mathcal{A}}_{\widetilde{W}}$ that can be analytically continued to P_∞ and vanish there by $\tilde{\mathcal{A}}^{reg}_{\widetilde{W}}$. Let $\tilde{\mathcal{V}}^{reg}_{\widetilde{W}}$ denote the same for $\mathcal{V}_{\widetilde{W}}$.

Definition 3.9. For a given vector field X in a neighborhood of the point $\tau_0 \in \widetilde{W}$, by $(\partial_X A_\tau)(z)$ we mean a full derivative of the function $\tilde{A}_\tau(d_\tau^{-1}(w))$ along the vector field X, for fixed w, where the differentiation is followed by the substitution $w = d_\tau(z)$:

$$\partial_X A_\tau(z) = \partial_X\left(\tilde{A}_\tau\left(d_\tau^{-1}(w)\right)\right)\Big|_{w=d_\tau(z)}.$$

Thus $\partial_X A_\tau$ is a local function of the variable z. We actually differentiate in moduli parameters the local function given by the restriction of the Krichever–Novikov function on the punctured neighborhood of the point P_∞. It is considered to be a function of the variable z, taking into account both the direct dependence on τ and the dependence via the gluing function between the local coordinate z, and the fixed local coordinate w. After the substitution $w = d_\tau(z)$ we regard $\partial_X A_\tau$ as a section of the sheaf $\tilde{\mathcal{A}}_{\widetilde{W}}$. Thus we have defined differentiation not only for Krichever–Novikov functions but for the sections of the sheaf of local functions as well.

For a fixed τ, the restriction $A_\tau|_{U_\infty}$ may be regarded as a local analytic function with the only singularity at P_∞. Let us ask how the differentiation along the vector field X changes the character of that singularity. We answer this question in terms of the Kodaira–Spencer cocycle $\rho(X)$ defined in Section 3.1. We assume below that $\rho(X) \in \mathcal{V}$, as asserted in the remark to Corollary 3.5. We can do this without loss of generality. Moreover, Corollary 3.5 asserts that the vector field $\rho(X)$ can be chosen in any given complement $\mathcal{V}_{(0)}$ to the kernel of the mapping Θ. As can be observed in the proof of Proposition 3.3, this ambiguity is related to the ambiguity in choosing an appropriate family of gluing functions d_τ. As soon as we have fixed $\mathcal{V}_{(0)}$, there is no longer any ambiguity in giving the $\rho(X)$, and the latter is completely given by the vector field X. Observe that the definition of $\partial_X A_\tau$ also depends on the choice of gluing functions. Here and below, we assume that the same family of gluing functions is chosen for the definition both $\rho(X)$ and $\partial_X A_\tau$.

Proposition 3.10. *For any section A of the sheaf $\mathcal{A}_{\widetilde{W}}$, and any local vector field X, define A^X by means the relation*

$$\partial_X A = -\rho(X)A + A^X \tag{3.3.1}$$

where A^X is the section of $\tilde{\mathcal{A}}_{\widetilde{W}}$. If A is the section of $\mathcal{A}^{reg}_{\widetilde{W}}$ then A^X is the section of $\tilde{\mathcal{A}}^{reg}_{\widetilde{W}}$. It is also true if A is the section of $\tilde{\mathcal{A}}^{reg}_{\widetilde{W}}$.

Proof. According to Definition 3.9, let us first compute a full derivative of the function $(\tilde{A}_\tau \cdot d_\tau^{-1})(w)$ along X, assuming w to be constant. Here the symbol $(\cdot)$ denotes the composition of maps.

By the formula of differentiation of the composition of maps (the chain rule) we have

$$\partial_X(\tilde{A}_\tau \cdot d_\tau^{-1}) = \tilde{\partial}_X \tilde{A}_\tau + \partial_z \tilde{A}_\tau \cdot \partial_X d_\tau^{-1}, \tag{3.3.2}$$

where $\tilde{\partial}_X \tilde{A}_\tau$ is a derivative along X, and, assuming the independence of z on τ, $\partial_z \tilde{A}_\tau$ is the differential of $\tilde{A}_\tau$ (in z-representation). Then, $\partial_X d_\tau^{-1} = -d_\tau^{-1} \cdot \partial_X d_\tau \cdot d_\tau^{-1}$, hence $\partial_z \tilde{A}_\tau \cdot \partial_X d_\tau^{-1} = \partial_z \tilde{A}_\tau \cdot (-d_\tau^{-1} \cdot \partial_X d_\tau) \cdot d_\tau^{-1}$. The expression in brackets is equal to $-\rho(X)$ by definition. Since the differential evaluated at a vector field gives the operator of derivation along that vector field, we obtain

$$\partial_z \tilde{A}_\tau \cdot \big(-d_\tau^{-1} \cdot \partial_X d_\tau\big) \cdot d_\tau^{-1} = \left(-\rho_z(X)\frac{\partial}{\partial z}\right) \cdot d_\tau^{-1}.$$

(Since the differential was in z-representation, we have obtained the differential operator in the variable z.)

By definition, we must do the substitution $w = d_\tau(z)$, which descends to erasing the d_τ^{-1} on the right-hand side. Observe that the first summand in (3.3.2) is calculated in the variable z from the very beginning. We obtain the following result:

$$\partial_X A_\tau(z) = \tilde{\partial}_X \tilde{A}_\tau(z) - \rho_z(X)\frac{\partial}{\partial z}\tilde{A}_\tau(z)$$

which we write as

$$\partial_X A_\tau = -\rho(X)A_\tau + \tilde{\partial}_X \tilde{A}_\tau \tag{3.3.3}$$

in coordinate free form in U.

Assume that $A_\tau \in \mathcal{A}^{reg}$ for any τ, i.e., $A_\tau(P_\infty) = 0$ where $P_\infty \in \Sigma(\tau)$, and $\tau \sim \tau_0$ is arbitrary. Since $z(P_\infty) = 0$, $A_\tau(P_\infty) = 0$ implies $\tilde{A}_\tau(0) = 0$ for all τ, hence $\tilde{\partial}_X \tilde{A}_\tau(0) = 0$. Set $A_\tau^X = \tilde{\partial}_X \tilde{A}_\tau$. Then the relation (3.3.3) results in the desired relation (3.3.1), where $A_\tau^X(0) = 0$ for any $\tau \sim \tau_0$, i.e., $A^X \in \tilde{\mathcal{A}}_{\widetilde{W}}^{reg}$, which completes the proof.

The last assertion in Proposition 3.10 is obvious. □

Let us now consider the sheaf $\mathcal{V}_{\widetilde{W}}$ of the vector-field Krichever–Novikov algebras. Let e be its meromorphic section. We could define $\partial_X e$ making use of the same argument as for functions, but we choose a shorter way here: the definition by the Leibniz rule. For any two sections $e \in \mathcal{V}_{\widetilde{W}}$, $A \in \mathcal{A}_{\widetilde{W}}$ by eA we mean the section obtained as a result of the action of the vector field on function over each point of the domain in question, and set by definition

$$(\partial_X e)A = \partial_X(eA) - e\partial_X A.$$

Proposition 3.11. *For any section e of the sheaf $\mathcal{V}_{\widetilde{W}}$, and any local vector field X define e^X by the relation*

$$\partial_X e = -[\rho(X), e] + e^X \tag{3.3.4}$$

where e^X is the section of $\mathcal{V}_{\widetilde{W}}$. If e is the section of $\mathcal{V}_{\widetilde{W}}^{reg}$, then e^X is the section of $\tilde{\mathcal{V}}_{\widetilde{W}}^{reg}$.

Proof. By Proposition 3.10 we have

$$\partial_X(eA) = -\rho(X)(eA) + (eA)^X,$$

where $(eA)^X = \tilde{\partial}_X(\widetilde{eA})$. The latter follows from the proof of Proposition 3.10. Similarly, $e(\partial_X A) = e(-\rho(X)A + \tilde{\partial}_X \tilde{A})$. Altogether, this implies

$$(\partial_X e)A = -[\rho(X), e]A + \tilde{\partial}_X(\widetilde{eA}) - e\tilde{\partial}_X \tilde{A}. \tag{3.3.5}$$

Recall that here $\tilde{\partial}_X$ means the derivation under the assumption of the independence of the local coordinate z of τ. In other words, it is a derivation of free sheaves (direct products) whose fibers are spaces of local functions or local vector fields, respectively.

Since the objects e and A are global, we have $\widetilde{eA} = \tilde{e}\tilde{A}$. Applying the Leibniz rule again, we obtain $\tilde{\partial}_X(\widetilde{eA}) - e\tilde{\partial}_X \tilde{A} = (\tilde{\partial}_X \tilde{e})\tilde{A}$. Hence (3.3.5) implies (3.3.4), where $e^X = \tilde{\partial}_X \tilde{e}$.

Let e be a section of the sheaf $\mathcal{V}_{\widetilde{W}}^{reg}$. Then $\tilde{\partial}_X \tilde{e}(0) = 0$ for the same reason as $\tilde{\partial}_X \tilde{A}(0) = 0$ in the proof of Proposition 3.10. Hence $e^X \in \tilde{\mathcal{V}}_{\widetilde{W}}^{reg}$; this completes the proof. □

To conclude, let us define $\partial_X B$ where B is a local section of a sheaf of operators acting on the sections of the $V_{\widetilde{W}}$. We set by definition $\partial_X B = [\partial_X, B]$, where ∂_X on the right-hand side is the operator of derivation of sections of the free sheaf $V_{\widetilde{W}}$. The ∂_X does not need any additional definition.

3.4 Projective flat connection and generalized Knizhnik–Zamolodchikov equation

In this section, we introduce a connection in the sheaf of conformal blocks (3.2.4), prove its projective flatness, and formulate our generalization of the Knizhnik–Zamolodchikov equations on the Riemann surfaces of positive genus with marked points.

For a local vector field X on $\widetilde{W}$, let us define the following first-order differential operator on sections of the free sheaf $V_{\widetilde{W}}$:

$$\nabla_X = \partial_X + T\big(\rho(X)\big) \tag{3.4.1}$$

where $\partial_X = \sum_i X_i(\tau)\frac{\partial}{\partial \tau_i}$ is a derivation of sections of the free sheaf $V_{\widetilde{W}}$, T is the Sugawara representation, $\rho(X)$ is the Kodaira–Spencer cocycle corresponding to the vector field X, and $\rho(X) \in \mathcal{V}$, as above. Below, passing to a certain quotient operator of ∇_X, we define an analogue of the Knizhnik–Zamolodchikov connection on the space $\mathcal{M}^{(1,0)}_{g,N+1}$.

Remark 3.12. With relation to the Knizhnik–Zamolodchikov equation, the Sugawara representation T appeared in [19] by those authors. However, the connection was not completely given there. In [74] (see also [77]) the connection was given in a maximally general form in terms of the Virasoro algebra. Being specialized to the rational case, when the role of $\mathcal{M}^{(1,0)}_{g,N+1}$ is played by the direct product of N Riemann spheres, this form descends to $\nabla_{\frac{\partial}{\partial z_i}} = \frac{\partial}{\partial z_i} + T(\frac{\partial}{\partial z_i})$, where z_i is a local coordinate on the i-th sphere (see [7, Introduction]). In this case, $\frac{\partial}{\partial z_i}$ is a tangent vector to the product of the spheres, but at the same time $\frac{\partial}{\partial z_i}$ is an element of the Virasoro algebra, and as such is associated with a certain Sugawara operator. For an arbitrary genus, there is no canonical way to associate X with any element of the Virasoro algebra. In this case they take an element e of the Virasoro algebra such that $X = \gamma(e)$ in [74] (the mapping γ was defined above in Section 3.1), and define the connection as $\nabla_X = \partial_X + T(\oplus_{i=1}^N e_i)$, where e_i is the Laurent series of the vector field e at the point P_i. This is already close to our formula (3.4.1) but the final step, namely replacing the $\oplus_{i=1}^N e_i$ with $\rho(X)$, is not done. The reason is as follows: $\rho(X)$ is a local vector field, but the Virasoro algebra does not provide any way of dealing with local vector fields, but only with formal ones which are objects of a different nature. The only way to work with local vector fields by means of the Virasoro algebra is to take their Laurent expansions in the neighborhoods of $P_1, \dots, P_N$, i.e., to reduce them to the Virasoro elements of the form $\oplus_{i=1}^N e_i$. The Krichever–Novikov bases provide a tool for a direct dealing with local vector fields. According to [27, Theorems 1.2 and 1.3] each local holomorphic vector field expands into a series over the elements of the Krichever–Novikov base. Thus its Sugawara representation makes sense. This enables us to regard $\rho(X)$ as an argument in the Sugawara representation and write the connection ∇ in the form (3.4.1). Moreover, as we have already mentioned, based on Corollary 3.5 $\rho(X)$ may be considered as an element of the Krichever–Novikov vector field algebra.

Remark 3.13. Indeed, there is no problem to continue by linearity the representation of operators onto local objects, whether they be functions or vector fields. It is more problematic for the multiplicative operations including multiplication of functions, commutator of vector fields, action of vector fields on functions and the corresponding actions on the elements of a representation space. This question is easy resolvable since we consider only almost-graded operations. Let us consider e.g. the expression eA, where e is a local vector field and A is a local function. If one of those objects belongs to the corresponding Krichever–Novikov algebra, i.e., expands over a finite number of basis vectors, or they are both infinite in the same direction (positive or negative with respect to the almost grading), then the result of the operation is well defined as an infinite series over the basis functions. Indeed, only a finite number of terms of the expansion for e, as well as for A, contributes into the component of a given degree of the result.

Proposition 3.14. *∇_X is well defined on conformal blocks.*

We call the connection defined by ∇_X on the sheaf of conformal blocks by *Knizhnik–Zamolodchikov connection.*

First, let us prove the following lemma.

Lemma 3.15. *If u is the fermion representation, then*

$$\partial_X u(A) = u(\partial_X A). \tag{3.4.2}$$

Proof. For any $u \in \mathfrak{g}$, $A \in \mathcal{A}$, and every basis fermion $\Psi = \psi_{i_1} \wedge \psi_{i_2} \dots$ we have $u(A)\Psi = (uA)\psi_{i_1} \wedge \psi_{i_2} \dots + \psi_{i_1} \wedge (uA)\psi_{i_2} \dots + \lambda_1 \cdot \Psi$, where the last term arises due to the regularization (page 27), and, for the same reason, in each of the expressions $(uA)\psi_{i_k}$ the term containing ψ_{i_k} is ignored (if there are some). By definition, $\partial_X u(A) = \partial_X \cdot u(A) - u(A)\partial_X$. Since the basis fermions do not depend on moduli, i.e., $\partial_X \Psi = 0$, we have

$$\begin{aligned}
(\partial_X u(A))\Psi &= \partial_X (u(A)\Psi) \\
&= (u\partial_X A)\psi_{i_1} \wedge \psi_{i_2} \dots + \psi_{i_1} \wedge (u\partial_X A)\psi_{i_2} \dots + (\partial_X \lambda_1) \cdot \Psi \\
&= u(\partial_X A)\Psi + (\partial_X \lambda_1) \cdot \Psi - \lambda_2 \cdot \Psi,
\end{aligned}$$

where $\lambda_2 \cdot \Psi$ arises due to regularization of $u(\partial_X A)$. If there is no need of any regularization, the relation (3.4.2) immediately follows. The regularization can be easily calculated in terms of the matrix of the operator uA in the space of sections of the holomorphic bundle related to our fermion representation. Set $uA = \sum_{i,j=-\infty}^{\infty} a_{ij} E_{ij}$ where $\{E_{ij} | i, j \in \mathbb{Z}\}$ is a natural base in the space of matrices. According to the regularization procedure,

$$\lambda_1 = \sum_{i \in N_-} a_{ii} - \sum_{i \in N_+} a_{ii},$$

where N_+ is the set of nonoccupied positions, or gaps, of a positive degree in the monomial Ψ, i.e., $N_+ = \mathbb{N} \setminus \{i_1, i_2, \cdots\}$, and N_- is a set of the occupied positions of negative or zero degree. Similarly, $\lambda_2 = \sum_{i \in N_-} \partial_X a_{ii} - \sum_{i \in N_+} \partial_X a_{ii}$, hence $\partial_X \lambda_1 - \lambda_2 = 0$, and the assertion is also true in the case of presence of regularization. □

While considering other classes of representations, we regard (3.4.2) as an additional requirement.

Proof of Proposition 3.14. Let A be a section of the sheaf $\tilde{\mathcal{A}}_{\widetilde{W}}$. By Lemma 3.15, and Theorems 2.21(ii) and 2.23, we obtain

$$\begin{aligned}\left[\nabla_X, u(A)\right] &= \left[\partial_X + T(\rho(X)), u(A)\right] = \left[\partial_X, u(A)\right] + \left[T\big(\rho(X)\big), u(A)\right] \\ &= u(\partial_X A) + u\big(\rho(X)A\big).\end{aligned}$$

Over each point of the moduli space A is a local function, but $\rho(X)$ is a Krichever–Novikov vector field; hence, according to the remark on page 68, all commutators in this chain of equations are well defined.

Assume that $A \in \tilde{\mathcal{A}}_{\widetilde{W}}^{reg}$. Then, according to Proposition 3.10 we have

$$u(\partial_X A) = -u\big(\rho(X)A\big) + u(A^X)$$

where $A^X \in \tilde{\mathcal{A}}_{\widetilde{W}}^{reg}$. Hence $[\nabla_X, u(A)] = u(A^X)$ and

$$\left[\nabla_X, u(A)\right] \in u\big(\tilde{\mathcal{A}}_{\widetilde{W}}^{reg}\big)$$

which imply

$$\left[\nabla_X, u(\tilde{\mathcal{A}}_{\widetilde{W}}^{reg})\right] \subseteq u\big(\tilde{\mathcal{A}}_{\widetilde{W}}^{reg}\big). \tag{3.4.3}$$

According to (3.2.4), the coinvariants are defined as $C = V/\mathfrak{g}^{reg} V$. Let $\tilde{V}$ be the completed module, in the sense that the infinite to the left combinations of the basis fermions are allowed. It is clear that $V/\mathfrak{g}^{reg} V = \tilde{V}/\tilde{\mathfrak{g}}^{reg} \tilde{V}$ (both $\tilde{\mathfrak{g}}^{reg}$, and $\tilde{V}$ consist of the series infinite in the same direction; hence the action is well-defined).

The relation (3.4.3) shows that $\mathfrak{g}^{reg} V_{\widetilde{W}}$ is a ∇_X-invariant subsheaf, hence ∇_X is well defined on $\tilde{V}_{\widetilde{W}}/\mathfrak{g}^{reg} \tilde{V}_{\widetilde{W}}$. □

Lemma 3.16. *Let u be a fermion, or an irreducible representation. Then for any local vector field X on $\widetilde{W}$ and for any section e of the sheaf $\mathcal{V}$ we have*

$$\partial_X T(e) = T(\partial_X e) + \lambda \cdot id,$$

where $\lambda = \lambda(X, e) \in \mathbb{C}$.

Proof. By virtue of the fundamental relation of Theorem 2.21(ii), for any $e \in \mathcal{V}$, $u \in \mathfrak{g}$, $A \in \mathcal{A}$

$$\big[T(e), u(A)\big] = u(eA). \tag{3.4.4}$$

Let us take the derivative of both sides of the relation (3.4.4) along the vector field X. By Lemma 3.15 we will obtain

$$\big[\partial_X T(e), u(A)\big] + \big[T(e), \partial_X u(A)\big] = u\big((\partial_X e)A\big) + u\big(e(\partial_X A)\big). \tag{3.4.5}$$

Again by (3.4.4), and by Lemma 3.15 the second terms on both hand sides of the relation (3.4.5) are equal. Hence

$$\big[\partial_X T(e), u(A)\big] = u\big((\partial_X e)A\big).$$

Applying (3.4.4) once more, let us replace the right-hand side of the last relation with the expression $[T(\partial_X e), u(A)]$ (see remark below). Thus,

$$\big[\partial_X T(e) - T(\partial_X e), u(A)\big] = 0 \tag{3.4.6}$$

for any $X \in \mathrm{T}\widetilde{W}$, $u \in \mathfrak{g}$, $A \in \mathcal{A}$.

For the representations u in question (listed in the lemma), commutativity of an operator with all representation operators implies that the operator is scalar. For an irreducible representationm this statement is simply the Schur lemma; for a fermion representation it follows from the uniqueness of the vacuum vector (indeed, there is an intersection between those classes of representations). □

Remark 3.17. $\partial_X e$ is a local vector field, or more precisely, the section of the sheaf $\tilde{\mathcal{V}}$. Hence, to prove (3.4.6) we need a relation (3.4.4) for local vector fields. By Remark 3.13, the representations $T(e)$, and $u(A)$ are also well defined for local vector fields and functions, respectively, and the relation (3.4.4) remains true. Indeed, for $A \in \mathcal{A}$, a homogeneous element $v \in V$ and an arbitrary n, there exists a partial sum $\tilde{e}$ of the vector field $\partial_X e$ expansion such that $(u((\partial_X e)A)v)_n = (u(\tilde{e}A)v)_n$, and $([T(\partial_X e), u(A)]v)_n = ([T(\tilde{e}), u(A)]v)_n$, where $(\cdot)_n$ denotes the projection onto the degree n component in $\tilde{V}$. By virtue of (3.4.4), the right-hand sides of the two last relations are equal, and thus their left-hand sides are also equal. This implies (3.4.4) with $\partial_X e$ instead e.

In the the remainder of this section we will prove a projective flatness of the connection (3.4.1) and introduce the corresponding analogue of the Knizhnik–Zamolodchikov equations.

Theorem 3.18. ∇_X *is a projective flat connection in the bundle of conformal blocks.*

Proof.

$$\begin{aligned}[\nabla_X, \nabla_Y] &= \Big[\partial_X + T\big(\rho(X)\big), \partial_Y + T\big(\rho(Y)\big)\Big] \\ &= [\partial_X, \partial_Y] + \Big[\partial_X, T\big(\rho(Y)\big)\Big] - \Big[\partial_Y, T\big(\rho(X)\big)\Big] \\ &\quad + \Big[T\big(\rho(X)\big), T\big(\rho(Y)\big)\Big].\end{aligned} \tag{3.4.7}$$

Since T is a projective representation of the Lie algebra $\mathcal{V}$, and in virtue of the relations $[\partial_X, T(\rho(Y))] = \partial_X T(\rho(Y))$, $[\partial_Y, T(\rho(X))] = \partial_Y T(\rho(X))$, we may write (3.4.7) in the form

$$[\nabla_X, \nabla_Y] = \partial_{[X,Y]} + T\Big(\partial_X \rho(Y) - \partial_Y \rho(X) + \big[\rho(X), \rho(Y)\big]\Big) + \lambda \cdot id. \tag{3.4.8}$$

Here we also apply Lemma 3.16. According to [74, Lemma 1.3.8],

$$\partial_X \rho(Y) - \partial_Y \rho(X) + \big[\rho(X), \rho(Y)\big] = \rho\big([X, Y]\big). \tag{3.4.9}$$

This implies that

$$\begin{aligned}[\nabla_X, \nabla_Y] &= \partial_{[X,Y]} + T\Big(\rho\big([X, Y]\big)\Big) + \lambda \cdot id \\ &= \nabla_{[X,Y]} + \lambda \cdot id.\end{aligned} \tag{3.4.10}$$

□

We consider the following equations of the horizontal sections of the connection ∇_X as a generalization of the Knizhnik–Zamolodchikov equations:

$$\nabla_X \Psi = 0, \quad X \in H^0\big(\widetilde{W}, \mathrm{T}\mathcal{M}^{(1)}_{g,N+1}\big), \tag{3.4.11}$$

where Ψ is a section of the sheaf of conformal blocks. These equations are proposed in [53]. In particular, the operator ∇_X is explicitly computed there for $g = 0$, $g = 1$, and slightly different representation of the affine algebra (the computation works also in the case under consideration in this section).

To obtain finitely many equations here, let X range over the images of the basis elements of the subspace $\mathcal{V}_{(0)}$ introduced in Corollary 3.5.

3.5 Explicit form of the Knizhnik–Zamolodchikov equations for genus 0 and genus 1

In this section we give an explicit form of the above introduced generalized Knizhnik–Zamolodchikov equations for genus 0, and genus 1. In particular, we show that for genus 0 our equations descend to the conventional form of the Knizhnik–Zamolodchikov equations. The section can be skipped for the first reading.

Applying (2.7.4), and setting

$$l_k^{(n,p)(m,s)} = \frac{1}{2\pi i}\int_{C_\tau} \omega^{(n,p)}\omega^{(m,s)} v_k,$$

where $\{v_k\}$ is a base in $\mathcal{V}_{(0)}$ (see the end of the previous section), the equations (6.5.1), we write

$$\left(\partial_k - \frac{1}{c+\varkappa}\sum_{n,m,p,s} l_k^{(n,p)(m,s)} :u(n,p)u(m,s):\right)\Phi = 0. \tag{3.5.1}$$

3.5.1 Explicit form of the equations for $g = 0$

Let us derive the original Knizhnik–Zamolodchikov equations

$$\left(\nu\frac{\partial}{\partial z_i} - \sum_{j\neq i}\frac{t_i^a t_j^a}{z_i - z_j}\right)\Phi = 0\,, \qquad i = 1,\dots,N \tag{3.5.2}$$

[19, equation (3.21)] from (3.5.1), where $\nu \in \mathbb{C}$ in (3.5.2).

Let z_i $(i = 1,\dots,N)$ denote the N marked points on the Riemann sphere. Let us fix the reference point z_∞ at the ∞. We set

$$\alpha(i) := \prod_{\substack{l=1\\ l\neq i}}^{N}(z_i - z_l)^{-1}$$

for $i = 1,\dots,N$. For any set $\{z_i | i = 1,\dots,N\}$, the elements of the Krichever–Novikov base of degree (-1) for quadratic differentials and for vector fields are given by the relations

$$\Omega^i(z) := \Omega^{-1,i} = \frac{dz^2}{z - z_i}, \quad e_i(z) := e_{-1,i} = \left(\alpha(i)\prod_{j\neq i}(z - z_j)\right)\frac{\partial}{\partial z}, \tag{3.5.3}$$

respectively. The vector field e_i is equal to $\frac{\partial}{\partial z}$ at the point z_i and vanishes at all other z_j, $j \neq i$. Hence e_i corresponds to the basis direction ∂_i on the configuration space corresponding to shifting the z_i. On the other hand, e_i is exactly the same as the

Krichever–Novikov dual vector field to the quadratic differential Ω^i. This follows from the calculation of residues.

The coefficients $l_k^{(m,i)(n,j)}$ are given by the relation

$$l_k^{(m,i)(n,j)} = \frac{1}{2\pi \mathrm{i}} \oint_C \omega^{(m,i)} \omega^{(n,j)} e_k. \tag{3.5.4}$$

In the rational case, the expression for the basis tensors is obvious:

$$f_{n,p}^{\lambda}(z) = (z - z_p)^{n-\lambda} \Bigg(\prod_{\substack{i=1 \\ i \neq p}}^{N} (z - z_i) \Bigg)^{n-\lambda+1} \Bigg(\prod_{\substack{i=1 \\ i \neq p}}^{N} (z_p - z_i) \Bigg)^{-n+\lambda-1} dz^{\lambda}. \tag{3.5.5}$$

By (3.5.5) we have

$$\omega^{(m,i)} = \frac{\alpha(i)^{-m}\, dz}{(z - z_i) \prod_{s=1}^{N} (z - z_s)^m}. \tag{3.5.6}$$

Thus the integrand in (3.5.4) is equal to

$$\frac{\alpha(i)^{-m} \alpha(j)^{-n} \alpha(k)\, dz}{(z - z_i)(z - z_j)(z - z_k) \prod_{s=1}^{N} (z - z_s)^{m+n-1}}. \tag{3.5.7}$$

The coefficients (3.5.4) can be obtained by summation of residues at the points $z_1, \dots, z_N$ (or by taking the residue at the point z_∞ with the inverse sign). Observe that if $m + n \leq -2$ or $m + n = -1$, but $i = j = k$ does not hold, then all residues at the points $z_1, \dots, z_N$ vanish. If $m + n > 0$, then the residue at z_∞ vanishes. We consider here only the case when Φ in (3.5.1) is the vector being annihilated by the subalgebra $\widehat{\mathfrak{g}}_+$. Observe that originally Knizhnik–Zamolodchikov equations were obtained under the same assumption in [19]. But then, if $m + n = 0$, and $m, n \neq 0$, then either m, or n is positive, and hence $:u(m,i)u(n,j):\ \Phi = 0$, because the elements of positive degree will be found on the right. Nonzero coefficients in (3.5.1), for a given $i, k = 1, \dots, N$, $i \neq k$ are

$$\begin{aligned} l_k^{(0,i)(0,i)} &= \frac{\alpha(i)^{-1}\alpha(k)}{z_i - z_k}, & l_k^{(0,i)(0,k)} &= l_k^{(0,k)(0,i)} = \frac{1}{z_k - z_i}, \\ l_k^{(0,k)(0,k)} &= \sum_{i \neq k} \frac{1}{z_k - z_i}, & l_k^{(-1,k)(0,k)} &= l_k^{(0,k)(-1,k)} = \alpha(k)^2. \end{aligned} \tag{3.5.8}$$

By vertical vector fields we mean the ones vanishing at all z_i, hence shifting none of those points. In the remainder of this subsection, we will show that by adding some vertical vector fields to the basis vector fields e_k, passing to a certain quotient, and computing the structure constants with respect to this base, we can annihilate all coefficients except for those with $m = n = 0$, and $i \neq j$. Hence the equations (3.5.1) will have the following form in the modified basis:

$$\left(\partial_i - \frac{1}{c+\kappa}\sum_{j\neq i}\frac{:u(0,i)u(0,j): + :u(0,j)u(0,i):}{z_i - z_j}\right)\Phi = 0, \tag{3.5.9}$$

where $i = 1, \dots, N$.

Observe that the coefficients $l_k^{(0,i)(0,j)}$ $(i \neq j)$ can also be obtained from the relation

$$\begin{aligned}\omega^{0,i}(z)\omega^{0,j}(z) &= \frac{\mathrm{d}z^2}{(z-z_i)(z-z_j)} = \frac{1}{z_i - z_j}\left(\frac{\mathrm{d}z^2}{z-z_i} - \frac{\mathrm{d}z^2}{z-z_j}\right)\\ &= \frac{1}{z_i - z_j}(\Omega^i - \Omega^j)\ (i \neq j),\end{aligned}$$

which implies

$$l_k^{(0,i)(0,j)} = \begin{cases} 0, & k \neq i \text{ and } k \neq j\\ \frac{1}{z_k - z_j}, & k = i\\ -\frac{1}{z_i - z_k}, & k = j.\end{cases} \tag{3.5.10}$$

Let us modify the vector fields e_k, $k = 1, \dots, N$ in order to annihilate the contribution of the terms with $i = j$, $m = n = 0$, not affecting the coefficients with $i \neq j$. Set $\tilde{\Omega}^i = \omega^{0,i}\omega^{0,i} = \frac{dz^2}{(z-z_i)^2}$. Denote the pairing (1.2.2) giving the Krichever–Novikov duality between 2-differentials and vector fields by the angle brackets $\langle\cdot,\cdot\rangle$. Then

$$\langle\tilde{\Omega}^i, e_k\rangle = \frac{1}{z_k - z_i}\prod_{s\neq i,k}\frac{z_i - z_s}{z_k - z_s}, \quad \text{for } k \neq i, \tag{3.5.11}$$

and

$$\langle\tilde{\Omega}^k, e_k\rangle = \sum_{s\neq k}\frac{1}{z_k - z_s}. \tag{3.5.12}$$

Let us pass from the set of vector fields $\{e_k \mid k = 1, \dots, N\}$ to the set of vector fields $\{e'_k \mid k = 1, \dots, N\}$ where

$$e'_k := e_k + \sum_{i=1}^{N}\lambda_{ki}E_i, \qquad \lambda_{ki} \in \mathbb{C}, \tag{3.5.13}$$

and $\{E_s \mid s = 1, \dots, N\}$ are the Krichever–Novikov basis elements of zero degree for vector fields, i.e.,

$$E_i := e_{0,i} = (z - z_i)\prod_{s\neq i}\frac{(z-z_s)^2}{(z_i - z_s)^2}\frac{\partial}{\partial z}. \tag{3.5.14}$$

Lemma 3.19.

(a) *E_i is a vertical vector field* $(i = 1,\ldots,N)$;

(b) $\langle\tilde{\Omega}^i, E_j\rangle = \delta_{ij} \quad (i,j = 1,\ldots,N)$;

(c) $\langle\Omega^i, E_j\rangle = 0 \qquad (i,j = 1,\ldots,N)$.

Proof. (a) follows from the vector fields E_k vanishing at all points $\{z_s | s = 1,\ldots,N\}$.

(b) By definition

$$\langle\tilde{\Omega}^i, E_j\rangle = \sum_{p=1}^{N} \operatorname{res}_{z_p} \frac{(z-z_j)}{(z-z_i)^2} \prod_{s\neq j} \frac{(z-z_s)^2}{(z_j-z_s)^2}\, dz.$$

If $i \neq j$ then the 1-form on the right hand side of the last relation is holomorphic, and all its residues are equal to zero. If $i = j$ then all residues vanish except for that at z_i where it is equal to 1.

(c) All 1-forms $\Omega^i E_j, i, j = 1,\ldots,N$ are holomorphic, which proves the assertion (see also (1.3.1)). □

By Lemma 3.19(a) the vector fields $\{e'_k \mid k = 1,\ldots,N\}$ correspond to the basis infinitesimal deformations ∂_i, as well as the vector fields $\{e_k \mid k = 1,\ldots,N\}$. By Lemma 3.19(c) the coefficients $l_k^{(m,i)(n,j)}$ $(i \neq j)$ do not change under the replacement of the vector fields $\{e_k \mid k = 1,\ldots,N\}$ with $\{e'_k \mid k = 1,\ldots,N\}$.

Now let us compute the coefficients λ_{ki} in (3.5.13) such that the following condition would be satisfied:

$$\langle e'_k, \tilde{\Omega}^i\rangle = 0, \quad i,k = 1,\ldots,N. \tag{3.5.15}$$

This means that $l_k^{(0,i)(0,i)} = 0$ $(i = 1,\ldots,N)$, provided that those coefficients are computed with respect to the vector fields $\{e'_k \mid k = 1,\ldots,N\}$. Let us recall that $\tilde{\Omega}^i = \omega^{0,i}\omega^{0,i}$.

Equation (3.5.15) means

$$\left\langle e_k + \sum_{s=1}^{N} \lambda_{ks} E_s, \tilde{\Omega}^i \right\rangle = 0, \quad i,k = 1,\ldots,N. \tag{3.5.16}$$

By Lemma 3.19(b) this is equivalent to $\langle e_k, \tilde{\Omega}^i\rangle + \lambda_{ki} = 0$. Hence (3.5.15) holds if and only if $\lambda_{ki} = -\langle e_k, \tilde{\Omega}^i\rangle$.

Taking account of (3.5.11), (3.5.12), we derive the following explicit expressions for e'_k from (3.5.16):

$$e'_k = e_k + \sum_{i\neq k} \frac{\prod_{s=1}^{N}(z-z_s)^2}{z_k - z_i} \cdot \alpha(k) \left(\frac{\alpha(i)}{z-z_i} - \frac{\alpha(k)}{z-z_k} \right) \frac{\partial}{\partial z}$$

where $i, k = 1, \dots, N$. Observe that there may arise some nonvanishing coefficients with $m+n = 1$. Using again the assumption that $\widehat{\mathfrak{g}}_+$ annihilates Φ, and the normal ordering, we will obtain that the corresponding terms will not contribute to the resulting equation.

Finally, the term $l_k^{(0,k)(-1,k)}(: u(0,k)u(-1,k) : + : u(-1,k)u(0,k) :)$ in equation (3.5.1) can be eliminated either by taking into consideration only the zero degree terms, or by passing to the quotient sheaf $V/\mathfrak{g}^{reg} V$ of the conformal blocks (in the rational case $\mathfrak{g}^{reg}$ begins with the degree -1). With this goal, one must suppose that $:u(0,k)u(-1,k):=:u(-1,k)u(0,k):= u(-1,k)u(0,k)$. The standard normal ordering (2.7.2) possesses this property.

Finally, let V be the Verma module (2.5.4). Let us take Φ from $\mathfrak{g}$-module $V_\lambda = V_{\lambda_1} \otimes \dots \otimes V_{\lambda_N}$ (see Section 2.5). If we suppose that u_a is the self-dual basis of the Lie algebra $\mathfrak{g}$, take the standard normal ordering, and set $u_a(0,i)\Phi = t_i^a \Phi$, then

$$u_a(0,j)u_a(0,i)\Phi = t_j^a t_i^a \,\Phi = t_i^a t_j^a \,\Phi \quad \text{for } j \neq i$$

(t_i^a and t_j^a commute for $i \neq j$ since they act in the different components of the tensor product). Hence (3.5.9) gains the form (3.5.2) where $\nu = (c + \varkappa)/2$.

3.5.2 Explicit form of the equations for $g = 1$

The aim of this section is to obtain the explicit expressions for the coefficients of the Knizhnik–Zamolodchikov equations in terms of the Weierstrass σ-function. The necessary expressions for the basis Krichever–Novikov elements in the elliptic case are given in Section 3.6 below.

1. Let us take the following set of vector fields corresponding to the movements of the points $z_1, \dots, z_N$: $e_k(z) = A_{0,k}(z)\frac{\partial}{\partial z}$, $k = 1, \dots, N$ (see Section 3.6), or, in the explicit form,

$$e_k(z) = \prod_{s \neq k} \frac{\sigma(z - z_s)}{\sigma(z_k - z_s)} \cdot \frac{\sigma(z_k - z_0)^N}{\sigma(z - z_0)^N} \cdot \frac{\sigma(z + \sum_{s \neq k} z_s - N z_0)}{\sigma(\sum_{s=1}^N z_s - N z_0)} \frac{\partial}{\partial z}. \tag{3.5.17}$$

Here z_0 is a fixed reference point. Our first aim is to find the contribution of the terms of the form $l_k^{(m,i)(n,j)}$ $: u(m,i)u(n,j) :$ into the equations. Observe that $\omega^{m,i} = A_{-m-1,i}dz$ by virtue of the duality relations. Consider several cases. Similar to the case of genus 0, let Φ be an element of the representation space having been vanished by the subalgebra $\widehat{\mathfrak{g}}_+$.

1.1. $m \neq 0, n \neq 0$.
Let us define $\tilde{\omega}^{m,i}$, $\tilde{\omega}^{n,j}$, $\tilde{e}_k$ by means of the relations $\omega^{m,i} = \sigma(z - z_i)^{-1}\tilde{\omega}^{m,i}$, $\omega^{n,j} = \sigma(z-z_j)^{-1}\tilde{\omega}^{n,j}$, $e_k = \sigma(z-z_k)^{-1}\tilde{e}_k$. Then $\operatorname{ord}_{z_s} \tilde{\omega}^{m,i} = -m$, $\operatorname{ord}_{z_s} \tilde{\omega}^{n,j} = -n$, $\operatorname{ord}_{z_s} \tilde{e}_k = 1$ $(s = 1, \dots, N)$, $\operatorname{ord}_{z_0} \tilde{\omega}^{m,i} = Nm$, $\operatorname{ord}_{z_0} \tilde{\omega}^{n,j} = Nn$, $\operatorname{ord}_{z_0} \tilde{e}_k =$

$-N$. Set $\tilde{\omega}_k^{(m,i)(n,j)} := \tilde{\omega}^{m,i}\tilde{\omega}^{n,j}\tilde{e}_k$. Hence

$$\omega^{m,i}\omega^{n,j}e_k = \frac{\tilde{\omega}_k^{(m,i)(n,j)}}{\sigma(z-z_i)\sigma(z-z_j)\sigma(z-z_k)}, \tag{3.5.18}$$

and $\mathrm{ord}_{z_s}\tilde{\omega}_k^{(m,i)(n,j)} = -m-n+1$ for all $s = 1,\dots,N$. If $m+n \geq 1$, then (3.5.18) is holomorphic at the point z_0. If $m+n=0$, then either $m>0$, or $n>0$. In both cases $:u(m,i)u(n,j):\Phi = 0$. For the same reason, in the case $m+n=-1$ we have $m=0$, $n=-1$ (or vice versa), but in this section we assume $m \neq 0$, $n \neq 0$, thus this case does not arise. We obtain only the case $m+n \leq -2$ as the remainder. Then we have $\mathrm{ord}_{z_s}\tilde{\omega}_k^{(m,i)(n,j)} = -m-n+1 \geq 3$ $(s=1,\dots,N)$ for the numerator in (3.5.18), and hence the 1-form is holomorphic at all marked points, even if some of the points z_i, z_j, z_k coincide.

Thus the case in question does not contribute to the Knizhnik–Zamolodchikov equation.

1.2. $m=0$, $n \neq 0$ (or vice versa).
Set $\omega^{0,i} = \hat{\omega}^{0,i} - \sum_{s=1}^N \gamma_{i,s}\omega^{-1,s}$ where $\hat{\omega}^{0,i} = A'_{-1,i}dz$, $A'_{-1,i}$ are defined by means of the relations (3.6.4), and $\gamma_{i,s}$ by means of the relations (3.6.5) below. The order of sums on the right-hand side of the last relation, at each of the moving points, is defined by the order of its first summand (other terms have bigger orders). If $n>0$, then $:u(0,i)u(n,j):\Phi = 0$. Using the definition of e_k, and the relations $\omega^{m,i} = A_{-m-1,i}dz$, we obtain $\hat{\omega}^{0,i}\omega^{n,j}e_k = A'_{-1,i}A_{-1-n,j}A_{0,k}\,dz$. The expressions (3.5.17) and (3.6.7)–(3.6.9) imply the following: if $n \leq -2$ then $-n+1 \geq 3$, and $\hat{\omega}^{0,i}\omega^{n,j}e_k$ is holomorphic at any marked point. Since $n \neq 0$, we only considered the case $n=-1$. In this case $\hat{\omega}^{0,i}\omega^{n,j}e_k$ has at most simple poles at the marked points. This is true only if $i=j=k$. In particular, the terms with γ are holomorphic and do not contribute to the result. The calculation of residues gives

$$l_k^{(0,k)(-1,k)} = 1 \quad (k=1,\dots,N). \tag{3.5.19}$$

In the same way we obtain a symmetric expression $l_k^{(-1,k)(0,k)} = 1$.

1.3. $m=n=0$.
To find out $l_k^{(0,i)(0,j)}$ we must consider the 1-forms $\omega^{0,i}\omega^{0,j}e_k$ where $\omega^{0,i} = \hat{\omega}^{0,i} - \sum\limits_{s=1}^N \gamma_{is}\omega^{-1,s}$, $\omega^{0,j} = \hat{\omega}^{0,j} - \sum\limits_{r=1}^N \gamma_{jr}\omega^{-1,r}$. The terms $\omega^{-1,s}\omega^{-1,r}e_k$ are holomorphic, since all their cofactors are holomorphic. A term of the form $\omega^{-1,s}\hat{\omega}^{0,j}e_k$ can have a pole (which is necessarily simple) only if $s=j=k$, and is holomorphic otherwise. Thus

$$l_k^{(0,i)(0,j)} = \langle\hat{\omega}^{0,i}\hat{\omega}^{0,j}, e_k\rangle - (\gamma_{jk}\delta_i^k + \gamma_{ik}\delta_j^k)\langle\omega^{-1,k}\hat{\omega}^{0,k}, e_k\rangle\,.$$

The second scalar product was already found, and turned out to be equal to $l_k^{(0,k)(-1,k)}$ which is just equal to 1. For the first term we have $\hat{\omega}^{0,i}\hat{\omega}^{0,j}e_k = A'_{-1,i}A'_{-1,j}A_{0,k}\,dz$. In the case $i \neq j$, this 1-form has a residue at z_i if $i = k$, and at z_j if $j = k$ (if i, j, k are mutually different, then this 1-form is holomorphic). For $i = k \neq j$ we have

$$l_k^{(0,k)(0,j)} = \frac{1}{\sigma(z_k - z_j)} \frac{\sigma(z_k - w_1)\sigma(z_k - w_2)\sigma(z_j - z_0)}{\sigma(z_j - w_1)\sigma(z_j - w_2)\sigma(z_k - z_0)} - \gamma_{jk}, \qquad (3.5.20)$$

and a similar expression for $i \neq k = j$. In (3.5.20) we have $w_1 + w_2 = z_i + z_0$, and w_1 is the same as in â (3.6.9). Observe that in the case $i = k \neq j$, this coefficient is attached to the operator-valued factor $:u(0,k)u(0,j):$, and in the case $i \neq k = j$ to the factor $:u(0,i)u(0,k):$.

Just as in the case $g = 0$, by adding certain vector fields of degree 0 to e_k we can annihilate the contribution of $\omega^{0,i}\omega^{0,i}$. Thus we come to the following form of the Knizhnik–Zamolodchikov equations corresponding to the moving points ($k = 1, \dots, N$):

$$\begin{aligned} \partial_k \Phi - \frac{1}{c+\kappa} \sum_{i \neq k} l_k^{(0,k)(0,i)} \left(:u(0,k)u(0,i): + :u(0,i)u(0,k):\right) \Phi \\ - \frac{1}{c+\kappa} \sum_{i=1}^{N} \left(:u(0,i)u(-1,i): + :u(-1,i)u(0,i):\right) \Phi = 0. \end{aligned} \qquad (3.5.21)$$

2. Further on, let us consider the equations corresponding to the deformations of the complex structure (i.e., to the variation of modular parameters). Consider the vector field

$$e_0 = \sigma(z - E)^{N+1} \sigma(z - z_0)^{-1} \prod_{s=1}^{N} \sigma(z - z_s)^{-1} \frac{\partial}{\partial z},$$

where $E = (N+1)^{-1}(z_0 + z_1 + \dots + z_N)$.

The vector field e_0 has simple poles at all points $z_1, \dots, z_N, z_0$. The following lemma implies that the corresponding tangent vector on the moduli space is nontrivial, and does not depend on the tangent vectors corresponding to the vector fields $e_1, \dots, e_N$. Hence it corresponds to deformations of the complex structure. In the lemma, the vector fields are considered to be vertical with regard to the moduli space $\mathcal{M}_{g,N}$.

Lemma 3.20. *For the points $z_1, \dots, z_n, z_0$ in a generic position, the vector field e_0 cannot be represented as a linear combination of the vector fields $e_1, \dots, e_N$, and vertical vector fields.*

Proof. Since the vector fields $e_1, \dots, e_N$ are regular at the points $z_1, \dots, z_N$, their residues at z_0 vanish (it is possible to discuss residues of vector fields on elliptic curves). All vertical vector fields also have zero residues at z_0. It is obvious for those which are regular at z_0, but those which have zeroes at $z_1, \dots, z_N$ also have a vanishing residue at z_0 because the sum of all residues is equal to zero. In contrast, in a generic point of the moduli space, the vector field e_0 has a simple pole at z_0 with a nonvanishing residue. This completes the proof. □

Let us consider several cases, as above.

2.1. $m, n \neq 0$.
Define $\tilde{\omega}_0^{(m,i)(n,j)}$ from the reation

$$\omega^{m,i}\omega^{n,j}e_0 = \frac{\tilde{\omega}_0^{(m,i)(n,j)}}{\sigma(z-z_i)\sigma(z-z_j)}, \tag{3.5.22}$$

Then $\operatorname{ord}_{z_s} \tilde{\omega}_0^{(m,i)(n,j)} = -m-n-1$ $(s = 1, \dots, N)$. Since the order of the denominator in (3.5.22) can be two at the most, the 1-form (3.5.22) is holomorphic at the points $z_1, \dots, z_N$ only if $-m-n-1 \geq 2$, i.e., if $m+n \leq -3$. If $m > 0$, or $n > 0$, then $:u(m,i)u(n,j):\ \Phi = 0$. Thus, either $m+n = -1$, or $m+n = -2$. Since $m, n \neq 0$, the first case cannot take place, and the only possibility in the second case is $m = n = -1$. In this case $i = j$, otherwise (3.5.22) is also holomorphic. Using explicit formulas we obtain

$$l_0^{(-1,i)(-1,i)} = \sigma(z_i - E)^{N+1}\sigma(z_i - z_0)^{-1}\prod_{s\neq i}\sigma(z_i - z_s)^{-1} \tag{3.5.23}$$

for $i = 1, \dots, N$.

2.2. $m = 0, n \neq 0$ (or vice versa).
In this case, we must set $m = 0$ in (3.5.22). Then $\operatorname{ord}_{z_s} \tilde{\omega}^{(0,i)(n,j)} = -(n+1)$ $(s = 1, \dots, N)$. If $n > 0$, then $:u(0,i)u(n,j):\ \Phi = 0$. If $-(n+1) \geq 2$, i.e., $n \leq -3$, the 1-form (3.5.22) is holomorphic. Hence, we only have to consider the cases $n = -1$, and $n = -2$.

If $n = -1$ then $\omega^{0,i}\omega^{-1,j}e_0 = \hat{\omega}^{0,i}\omega^{-1,j}e_0 - \sum_{s=1}^{N}\gamma_{i,s}\omega^{-1,s}\omega^{-1,j}e_0$. If $s \neq j$ then the 1-forms $\omega^{-1,s}\omega^{-1,j}e_0$ are holomorphic at all moving points (see above). If $s = j$, then the contribution of those 1-forms is equal to $l_0^{(-1,j)(-1,j)}$ (see case 2.1 above). Moreover,

$$\hat{\omega}^{0,i}\omega^{-1,j}e_0 = \frac{F(z)}{\sigma(z-z_i)\sigma(z-z_j)}dz,$$

where

$$F(z) := A'_{-1,i}(z)A_{0,j}(z)\sigma(z-z_0)^{-1}\sigma(z-E)^{N+1}\sigma(z-z_i)\prod_{s\neq j}\sigma(z-z_s)^{-1}$$

and $A'_{-1,i}(z)$, $A_{0,j}(z)$ are introduced on page 82. As it follows from Section 3.6, $F(z)$ has the order 0 at the points $z_1, \dots, z_N$. If $i \neq j$, then there are two residues: at z_i, and at z_j. We thus obtain

$$l_0^{(0,i)(-1,j)} = \frac{F(z_i) - F(z_j)}{\sigma(z_i - z_j)} - \gamma_{ij} l_0^{(-1,j)(-1,j)} \quad \text{for} \quad i \neq j. \tag{3.5.24}$$

The contribution of this case to the equation is equal to $l_0^{(0,i)(-1,j)}$ $:u(0,i)u(-1,j):$. Moreover, $l_0^{(-1,i)(0,j)} = l_0^{(0,j)(-1,i)}$, but it contributes with the operator-valued coefficient $:u(-1,i)u(0,j):$. In the case $i = j$, there arise second-order poles, and

$$l_0^{(0,i)(-1,i)} = \frac{d}{dz}\left((z - z_i)^2 F(z)\right)\Big|_{z=z_i} - \gamma_{ii} l_0^{(-1,i)(-1,i)}.$$

Consider now the case $n = -2$. Then the order of the numerator in (3.5.22) is equal to 1 at all marked points. Thus, the residue can vanish only if $i = j$. The additional terms $\gamma_{is}\omega^{-1,s}\omega^{-2,i}e_0$ are holomorphic and will not contribute to the equation. By the explicit formulas we have

$$l_0^{(0,i)(-2,i)} = \sigma(z_i - E)^{N+1}\sigma(z_i - z_0)^{-1}\prod_{s\neq i}\sigma(z_i - z_s)^{-1}. \tag{3.5.25}$$

2.3. $m = n = 0$.
In this case

$$\omega^{0,i}\omega^{0,j}e_0 = \hat{\omega}^{0,i}\hat{\omega}^{0,j}e_0 - \sum_{s=1}^{N}\gamma_{is}\omega^{-1,s}\hat{\omega}^{0,j}e_0 - \sum_{r=1}^{N}\gamma_{jr}\omega^{-1,r}\hat{\omega}^{0,i}e_0$$
$$+ \sum_{s,r=1}^{N}\gamma_{is}\gamma_{jr}\omega^{-1,s}\omega^{-1,r}e_0.$$

The contributions of the terms containing γ can be expressed via the coefficients computed in cases 2.1 and 2.2. For the first term of the considered sum we have $\hat{\omega}^{0,i}\hat{\omega}^{0,j}e_0 = F_{ij}(z)\,dz$ where

$$F_{ij}(z) = C_{-1,i}^{-1}C_{-1,j}^{-1}\frac{\sigma(z - w_1)^2\sigma(z - w_2)^2}{\sigma(z - z_i)\sigma(z - z_j)}\prod_{s=1}^{N}\sigma(z - z_s)^{-1} \tag{3.5.26}$$
$$\times\,\sigma(z - E)^{N+1}\sigma(z - z_0)^{-3}, \tag{3.5.27}$$

and the coefficients are given by the relations (3.6.9). This term has second-order poles, and even third-order poles if $i = j$, at the points z_i, z_j. The corresponding coefficients are as follows:

$$l_0^{(0,i)(0,j)} = \tag{3.5.28}$$
$$\langle\hat{\omega}^{0,i}\hat{\omega}^{0,j}, e_0\rangle - \sum_{s=1}^{N}\left(\gamma_{is}l_0^{(0,j)(-1,s)} + \gamma_{js}l_0^{(0,i)(-1,s)} + \gamma_{is}\gamma_{js}l_0^{(-1,s)(-1,s)}\right),$$

where

$$\langle \hat{\omega}^{0,i}\hat{\omega}^{0,j}, e_0\rangle = \frac{d}{dz}\left((z-z_i)^2 F_{ij}(z)\right)\Big|_{z=z_i} + \frac{d}{dz}\left((z-z_j)^2 F_{ij}(z)\right)\Big|_{z=z_j} \quad (i \neq j),$$

$$\langle \hat{\omega}^{0,i}\hat{\omega}^{0,i}, e_0\rangle = \frac{1}{2}\frac{d^2}{dz^2}\left((z-z_i)^3 F_{ij}(z)\right)\Big|_{z=z_i}.$$

As for genus 0, all terms of the equations containing $u(n,j)$, $n < -1$ can be eliminated by passing to coinvariants, and we will obtain the complex structure deformation equations in the following form:

$$\partial_0\Phi - \frac{1}{c+\kappa}\sum_{i,j=1}^{N} l_0^{(0,i)(0,j)} :u(0,i)u(0,j):\Phi \tag{3.5.29}$$

$$-\frac{1}{c+\kappa}\sum_{i,j=1}^{N} l_0^{(0,i)(-1,j)}(:u(0,i)u(-1,j): + :u(-1,j)u(0,i):)\Phi \tag{3.5.30}$$

$$-\frac{1}{c+\kappa}\sum_{i=1}^{N} l_0^{(-1,i)(-1,i)} :u(-1,i)u(-1,i):\Phi \quad = \quad 0. \tag{3.5.31}$$

3.6 Appendix: the Krichever–Novikov base in the elliptic case

In the elliptic case (i.e., $g = 1$), the canonical bundle $\mathcal{K}$, and all its degrees, are trivial. For this reason this case is peculiar with respect to the conditions defining finitely many basis elements. In this section, following [45], we give explicit expressions for the basis elements in terms of the Weierstrass σ-function.

By triviality $\mathcal{K}$ we have the following expression for the basis λ-forms:

$$f^{\lambda}_{n,p} = A_{n-\lambda,p}dz^{\lambda}\,. \tag{3.6.1}$$

In particular, choosing a base in $\mathcal{F}^0$ we determine the base in $\mathcal{F}^{\lambda}$. A standard condition in the situation with N incoming, and one outgoing point ($P_1, P_2, \ldots, P_N$, and P_∞, respectively) is as follows:

$$\mathrm{ord}_{P_p}(A_{n,p}) = n, \quad \mathrm{ord}_{P_i}(A_{n,p}) = n+1 \text{ for } i \neq p, \ \mathrm{ord}_{P_\infty}(A_{n,p}) = -N(n+1)\,.$$

If, further on, $N = 1$, set $A_0 = 1$, and $A'_{-1}dz = \rho = (\omega^0)'$, as above. To satisfy the duality relations, we modify these elements. By setting

$$A_{-1} := A'_{-1} - \frac{c}{2}A_0 \tag{3.6.2}$$

where

$$c = \frac{1}{2\pi\mathrm{i}}\oint_C {A'_{-1}}^2 dz, \tag{3.6.3}$$

we obtain the desired relations, namely $\langle A_{-1}, \omega^0 \rangle = 0$. Observe that it is not necessary now for $\omega^0 = A_{-1}dz$ to have purely imaginary periods.

If $N > 1$, we impose the standard constrains on $A_{0,p}$, hence, also on $\omega^{-1,p}$. For $n = -1$ these constrains are modified as follows:

$$\mathrm{ord}_{P_p}(A'_{-1,p}) = -1, \quad \mathrm{ord}_{P_i}(A'_{-1,p}) = 0, \text{ for } i \neq p, \quad \mathrm{ord}_{P_\infty}(A'_{-1,p}) = -1. \tag{3.6.4}$$

By adding certain linear combinations of the functions $A_{0,s}$ to $A'_{-1,p}$, we obtain a basis again satisfying the duality condition. In more detail:

$$\gamma_{r,p} := \frac{1}{2}\frac{1}{2\pi\mathrm{i}}\oint_C A'_{-1,r}A'_{-1,p}dz. \tag{3.6.5}$$

The modified basis elements are given by the formula

$$A_{-1,r} := A'_{-1,r} - \sum_{s=1}^{N} \gamma_{r,s}A_{0,s}. \tag{3.6.6}$$

The explicit representations of the basis elements for $g = 1$ in terms of the Weierstrass σ-function are as follows [43].

Let the torus be given by $T = \mathbb{C}/L$ with a lattice $L = \langle 1, \tau \rangle_{\mathbb{Z}}$. Let $z_i \in \mathbb{C}$ be a pullback of the point $P_i \in T$ to $\mathbb{C}$, $i = 1, \dots, N$ (i.e., $z_i \bmod L = P_i$), and z_0 be a pullback of $P_\infty \in T$. For $n \in \mathbb{Z}$, $p = 1, \dots, N$ set

$$b_{n,p} = b_{n,p}(z_1, \dots, z_N, z_0) = -(n+1)\sum_{i=1}^{N} z_i + z_p + N(n+1)z_0.$$

If $n \neq -1$, and $N > 1$, or $n \neq 0, -1$, and $N = 1$, then $b_{n,p}$ is not congruent modulo L with z_i's, z_0 in general, even for small perturbations of the marked points. Let us choose $z_i = (z - z_i)$ as a local coordinate at P_i, $i = 0, 1, \dots, N$.

Proposition 3.21 ([45, p. 53]). *Let P_i, $i = 1, \dots, N$, and P_∞ be in a generic position. Then the basis elements $A_{n,p}$ for $n \neq 0, -1$, and $N = 1$, and for $n \neq -1$, and $N > 1$, are given by the relations*

$$A_{n,p}(z) = C_{n,p}^{-1}\left(\prod_{i=1}^{N}\sigma(z - z_i)\right)^{n+1}\sigma(z - z_p)^{-1}\sigma(z - z_0)^{-N(n+1)}\sigma(z - b_{n,p}), \tag{3.6.7}$$

where

$$C_{n,p} := \left(\prod_{i \neq p}\sigma(z_p - z_i)\right)^{n+1}\sigma(z_p - z_0)^{-N(n+1)}\sigma(z_p - b_{n,p}).$$

In particular, they are well-defined functions on the torus.

The point $b_{n,p}$ mod L is an additional zero of the basis element $A_{n,p}$. Observe that the constants $C_{n,p}$ also depend on the points $z_0, \dots, z_N$.

For the singular degrees the results are as follows. For $n = 0$: If $N = 1$, we set $A_0 = 1$. For $N > 1$ the above formula is still valid. It reads as

$$A_{0,p} = C_{0,p}^{-1}\Big(\prod_{i \neq p} \sigma(z - z_i)\Big)\sigma(z - z_0)^{-N}\sigma\Big(z + \sum_{i \neq p} z_i - Nz_0\Big), \tag{3.6.8}$$

where

$$C_{0,p} := \Big(\prod_{i \neq p} \sigma(z_p - z_i)\Big)\sigma(z_p - z_0)^{-N}\sigma\Big(z_p + \sum_{i \neq p} z_i - Nz_0\Big).$$

For $n = -1$: In this case the order at the point P_∞ is equal to -1. We must find out the elements $w_1, w_2 \in \mathbb{C}$ such that $w_1 + w_2 = z_p + z_0$ but $w_1, w_2 \neq z_i$ mod L, $i = 0, \dots, N$. The w_1 can be chosen arbitrarily; it determines w_2.

Proposition 3.22 ([45]). *In a generic position,*

$$A_{-1,p} = C_{-1,p}^{-1}\sigma(z - z_p)^{-1}\sigma(z - z_0)^{-1}\sigma(z - w_1)\sigma(z - (z_p + z_0 - \omega_1)) \tag{3.6.9}$$

where

$$C_{-1,p} := \sigma(z_p - z_0)^{-1}\sigma(z_p - w_1)\sigma(\omega_1 - z_0).$$

The ambiguity in the choice of w_1 originates in the fact that the function is determined by its orders at the points P_i's, and P_∞ up to an additive constant.

We have now $(\omega^{0,p})' = A'_{-1,p}dz$, and $\omega^{-1,p} = A_{0,p}dz$. Observe, nevertheless, that the duality relations still are not satisfied. In particular, the relation $\langle \omega^{0,p}, A'_{-1,s}\rangle = 0$ does not hold. To satisfy this constraint, one must add the same combination of elements $A_{0,r}$ to $A'_{-1,s}$ as in (3.6.5), (3.6.6).

Chapter 4

Lax operator algebras

In this chapter we begin studying one more class of almost-graded infinite-dimensional Lie algebras, referred to as *Lax operator algebras*. These algebras arose as a byproduct of the theory of integrable systems, and have already contributed to both the foundations of and new results of that theory. They are closely related to both the Kac–Moody and the Krichever–Novikov affine algebras. For example, they have the same associated graded object, but the relation between these three classes of Lie algebras is much more profound than this formal analogy.

In [33], I. Krichever invented a general concept of Lax operators on algebraic curves, generalizing the theory of conventional Lax and zero curvature equations with the *rational spectral parameter* on the case of algebraic curves Σ of arbitrary genus g. The linear space of such operators associated with fixed Tyurin data (Section 2.2), and an effective divisor $D = \sum_k n_k P_k$, $P_k \in \Sigma$ were defined as the space of meromorphic $(n \times n)$ matrix-valued functions on Σ having poles of multiplicity at most n_k at the points P_k, and at most simple poles at the ng points γ_s of the Tyurin divisor. The coefficients of the Laurent expansion of those matrix-valued functions in the neighborhood of a point γ_s had to obey certain linear constraints given in terms of the Tyurin parameters α_s (see relations (4.1.1)–(4.1.3) below). The concept of Lax operators with the spectral parameter on a Riemann surface arose as a result of long term study of the role of vector bundles in the theory of integrable systems [23,25,34]. It clarifies the reason for Tyurin parameters appearing in the context of Lax operators.

In a joint work [35] we observed that the space of such matrix-valued functions of arbitrary orders at the points P_k is closed with respect to the point-wise matrix multiplication. We constructed the orthogonal and symplectic analogues of Lax operators with the spectral parameter on a Riemann surface. In these cases, the Lax operators do not form any associative algebra, but only form Lie algebras. In terms of that work, the previous paper [33] treats the case of the Lax operators corresponding to $\mathfrak{gl}(n)$.

In this chapter, we define case by case the Lax operator algebras corresponding to the classical simple (and one reductive) Lie algebras. It would be a worthwhile task for a future researcher to give the corresponding theory in terms of the root systems, suitable for all types of simple algebras, including the singular ones. The problem is not purely algebraic. As the reader will see below, the analytic behavior of the Lax operators at the Tyurin points depends on the type of the root system. For example, its order is equal to 1 for all classical simple Lie algebras except for $\mathfrak{sp}(2n)$, but in the last case it is equal 2.

In Section 4.1 we prove the multiplicative properties of $\mathfrak{gl}(n)$-valued Lax operators, introduce $\mathfrak{g}$-valued Lax operators for $\mathfrak{g} = \mathfrak{so}(n)$ and $\mathfrak{g} = \mathfrak{sp}(2n)$, and prove that they are closed with respect to the pointwise bracket.

In Section 4.2, we define and study the almost-graded structure on Lax operator algebras.

In Section 4.3, we construct almost-graded central extensions of the Lax operator algebras. The corresponding 2-cocycles are quite similar to those for Krichever–Novikov, and Kac–Moody algebras. More precisely, they are distinct from them by a coboundary, but this coboundary makes the cocycles local.

Finally, in Section 4.4 we formulate the uniqueness theorem for almost-graded central extensions. A corresponding result was obtained in [54].

4.1 Lax operators and their Lie bracket

4.1.1 Lax operator algebras for $\mathfrak{gl}(n)$ and $\mathfrak{sl}(n)$

Every Lax operator algebra is given by a Riemann surface Σ with marked points $P_1, \dots, P_N$ [1] and the Tyurin data (see Section 2.2 for the latter). This is illustrated in the following figure:

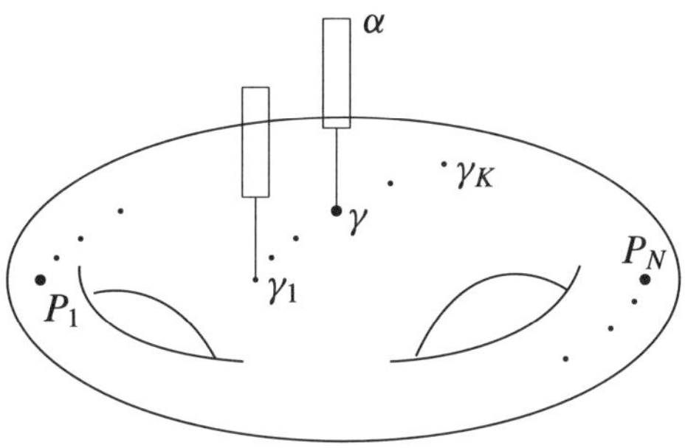

Figure 4.1. Tyurin data.

Given the Tyurin parameters $\{\alpha_s, \gamma_s | s = 1, \dots, gr\}$, and following the lines of [33], consider a $\mathfrak{gl}(n)$-valued function L on Σ, holomorphic outside $P_1, \dots, P_N$ and $\{\gamma_s | s = 1, \dots, gr\}$, and having at most simple poles at the points in the last set. It is called a *Lax operator* if

$$L = \frac{L_{s,-1}}{z - z_s} + L_{s0} + O(z - z_s), \quad z_s = z(\gamma_s), \tag{4.1.1}$$

[1] In considerations related to the almost-graded structure we distinguish one of the marked points and denote it as P_∞.

where z is a local coordinate in the neighborhood of a γ_s, z_s is the coordinate of the γ_s itself, and the following relations hold:

(i) $L_{s,-1} = \alpha_s \beta_s^t$ where β_s^t is a row, and

$$\operatorname{tr} L_{s,-1} = \beta_s^t \alpha_s = 0, \tag{4.1.2}$$

where $\alpha_s \in \mathbb{C}^n$ is fixed, $\beta_s \in \mathbb{C}^n$ is arbitrary, and the upper t denotes a transposed matrix. In particular, $L_{s,-1}$ has rank 1;

(ii) α_s is an eigenvector of the matrix L_{s0}:

$$L_{s0}\alpha_s = \varkappa_s \alpha_s. \tag{4.1.3}$$

Lemma 4.1. *Let L' and L'' satisfy the conditions* (4.1.1)–(4.1.3). *Then $L = L'L''$ also satisfies them.*

Proof. From (4.1.1) we have

$$L = \frac{L'_{s,-1}L''_{s,-1}}{(z-z_s)^2} + \frac{L'_{s,-1}L''_{s0} + L'_{s0}L''_{s,-1}}{(z-z_s)} + L'_{s,-1}L''_{s1} + L'_{s0}L''_{s0} + L'_{s1}L''_{s,-1} + O(1). \tag{4.1.4}$$

From (4.1.2) for L' it follows that the first term vanishes:

$$L'_{s,-1}L''_{s,-1} = \alpha_s(\beta_s'^t\alpha_s)\beta_s''^t = 0.$$

For the second one we have $L_{s,-1} = L'_{s,-1}L''_{s0} + L'_{s0}L''_{s,-1} = \alpha_s(\beta_s'^t L''_{s0}) + (L'_{s0}\alpha_s)\beta_s''^t$. By (4.1.2) for L' it follows $L'_{s0}\alpha_s = \varkappa'_s\alpha_s$, hence $L_{s,-1} = \alpha_s\beta_s^t$ where $\beta_s^t = \beta_s'^t L''_{s0} + \varkappa'_s\beta_s''^t$. Further, $\operatorname{tr} L_{s,-1} = (\beta_s'^t L''_{s0} + \varkappa'_s\beta_s''^t)\alpha_s = \varkappa''_s\beta_s'^t\alpha_s + \varkappa'_s\beta_s''^t\alpha_s = 0$.

Consider the expression $L_{s,0}\alpha_s$, where $L_{s,0} = L'_{s,-1}L''_{s1} + L'_{s0}L''_{s0} + L'_{s1}L''_{s,-1}$. From the definition of the Lax operators it follows that $L''_{s,-1}\alpha_s = 0$ and $L'_{s0}L''_{s0}\alpha_s = \varkappa'_s\varkappa''_s\alpha_s$. We also have $L'_{s,-1}L''_{s1}\alpha_s = \alpha_s(\beta_s'^t L''_{s1}\alpha_s)$. Hence α_s is an eigenvector of the matrix $L_{s,0}$ with the eigenvalue $\varkappa_s = \beta_s'^t L''_{s1}\alpha_s + \varkappa'_s\varkappa''_s$. □

Since the conditions (4.1.1) and (4.1.2) are linear, the Lax operators constitute an associative algebra, and thus the corresponding Lie algebra. The latter is called *Lax operator algebra*.

If besides (4.1.1)–(4.1.3) the function L satisfies the condition $\operatorname{tr} L = 0$, it is called *an $\mathfrak{sl}$-valued Lax operator*. Such Lax operators form a Lie algebra.

4.1.2 Lax operator algebras for $\mathfrak{so}(n)$

For the elements of $\mathfrak{g} = \mathfrak{so}(n)$ we have $X^t = -X$. We introduce the matrix-function L taking values in $\mathfrak{so}(n)$ by the same expansion as in Section 4.1.1 but changing the requirement (i) of that section for the reason that there is no rank one skew-symmetric matrix, and we then change accordingly the requirement (ii). We set $z_s = 0$, omit the lower s for brevity and write the expansion (4.1.1) in the form

$$L = \frac{L_{-1}}{z} + L_0 + O(z), \tag{4.1.5}$$

where $L_0, L_1, \ldots$ are skew-symmetric. Instead of condition (i) of Section 4.1.1 we require

$$L_{-1} = \alpha\beta^t - \beta\alpha^t, \tag{4.1.6}$$

where $\alpha \in \mathbb{C}^n$ is fixed, $\beta \in \mathbb{C}^n$ is arbitrary, and

$$\alpha^t\alpha = \beta^t\alpha (= \alpha^t\beta) = 0. \tag{4.1.7}$$

Like (4.1.3) we require

$$L_0\alpha = \varkappa\alpha \tag{4.1.8}$$

for some complex number $\varkappa$.

We will now prove that the space of Lax operators is also closed with respect to the point-wise Lie bracket in the case $\mathfrak{g} = \mathfrak{so}(n)$ in question. We stress that there is no associative algebra structure in this case.

Lemma 4.2. *The properties* (4.1.5)—(4.1.8) *are invariant with respect to the Lie bracket.*

Proof. (a) First, prove the absence of the term with z^{-2}. The corresponding coefficient equals

$$\begin{aligned}[L_{-1}, L'_{-1}] &= [\alpha\beta^t - \beta\alpha^t, \alpha\beta'^t - \beta'\alpha^t] \\ &= (\beta^t\alpha)(\alpha\beta'^t - \beta'\alpha^t) - (\alpha^t\alpha)(\beta'\beta^t - \beta\beta'^t) - (\alpha^t\beta')(\alpha\beta^t - \beta\alpha^t).\end{aligned}$$

It vanishes due to (4.1.7) (applied to both β and β'). We stress that for the product $L_{-1}L'_{-1}$ the term with z^{-2} does not vanish.

(b) Now calculate the term with z^{-1} in the product LL'. The coefficient is equal to

$$\begin{aligned}L_{-1}L'_0 + L_0L'_{-1} &= \alpha(\beta^tL'_0) - \beta(\alpha^tL'_0) + (L_0\alpha)\beta'^t - (L_0\beta')\alpha^t \\ &= \alpha(\beta^tL'_0) - \beta(-\varkappa'\alpha^t) + \varkappa\alpha\beta'^t - (L_0\beta')\alpha^t \\ &= \alpha(\beta^tL'_0 + \varkappa\beta'^t) - (L_0\beta' - \varkappa'\beta)\alpha^t\end{aligned}$$

(here we used the relation (4.1.8)). We see that it is not of the required form (4.1.6). Now consider the corresponding term of the commutator of L and L'. Then we obtain

$$\begin{aligned}[L, L']_0 &= \alpha(\beta^tL'_0 - \beta'^tL_0 + \varkappa\beta'^t - \varkappa'\beta^t) - (L_0\beta' - L'_0\beta - \varkappa'\beta + \varkappa\beta')\alpha^t \\ &= \alpha\beta''^t - \beta''\alpha^t\end{aligned}$$

where $\beta''^t = \beta^tL'_0 - \beta'^tL_0 + \varkappa\beta'^t - \varkappa'\beta^t$. It is easy to check that β'' satisfies (4.1.7).

(c) Check the requirement (4.1.8) on the eigenvalue of the degree 0 matrix coefficient $(LL')_0 = L_{-1}L'_1 + L_0L'_0 + L_1L'_{-1}$.

By (4.1.7) $L'_{-1}\alpha = 0$, hence for the third term we have $L_1L'_{-1}\alpha = 0$.

For the first term we have

$$L_{-1}L'_1\alpha = (\alpha\beta^t - \beta\alpha^t)L'_1\alpha = \alpha(\beta^t L'_1\alpha) - \beta(\alpha^t L'_1\alpha).$$

The last term on the right hand side of this equation vanishes due to skew-symmetry of L'_1.

Thus

$$(LL')_0\alpha = \varkappa''\alpha \quad \text{where} \quad \varkappa'' = \beta^t L'_1\alpha + \varkappa\varkappa'. \tag{4.1.9}$$

□

4.1.3 Lax operator algebras for $\mathfrak{sp}(2n)$

For the elements of the symplectic algebra we have $X^t = -\sigma X\sigma^{-1}$ where σ is a nondegenerate skew-symmetric matrix defining the algebra.

Take the expansion for L in the form

$$L = \frac{L_{-2}}{z^2} + \frac{L_{-1}}{z} + L_0 + L_1z + L_2z^2 + O(z^3) \tag{4.1.10}$$

(again, we omit the index s for brevity), where $L_{-2}, L_{-1}, L_0, L_1, \dots$ are symplectic matrices, and

$$L_{-2} = \nu\alpha\alpha^t, \;\; L_{-1} = (\alpha\beta^t + \beta\alpha^t)\sigma \quad (\nu \in \mathbb{C},\; \beta \in \mathbb{C}^{2n}). \tag{4.1.11}$$

Just as in (4.1.7), we require

$$\beta^t\sigma\alpha = 0. \tag{4.1.12}$$

Notice that $\alpha^t\sigma\alpha = 0$, due to the skew-symmetry of the matrix σ.

Now we require

$$L_0\alpha = \varkappa\alpha \tag{4.1.13}$$

for some complex number $\varkappa$.

We impose a new relation

$$\alpha^t\sigma L_1\alpha = 0. \tag{4.1.14}$$

Prove now that the space of Lax operators is closed with respect to the Lie bracket in the case $\mathfrak{g} = \mathfrak{sp}(2n)$ in question. We stress that in this case there is also no structure of associative algebra.

Lemma 4.3. *The properties* (4.1.10)–(4.1.14) *are invariant with respect to the Lie bracket.*

Proof. Let $L'' = [L, L']$.

(a) The absence of the terms of orders -4 and -3 in z in L'' follows only from the relations $\beta^t\sigma\alpha = 0$, $\alpha^t\sigma\alpha = 0$ (4.1.12).

(b) For the order -2-term we have

$$L''_{-2} = (\nu'\varkappa - \nu\varkappa' + \beta^t\sigma\beta')\alpha\alpha^t\sigma;$$

hence it has the form required by the relation (4.1.11) (here and below, ν', β', L'_i have the same meaning for L' as ν, β, L_i for L).

(c) The form of the order -1-term: a straightforward calculation using (4.1.10), (4.1.11) gives

$$\begin{aligned}L''_{-1} = \alpha\big(&\nu\cdot\alpha^t\sigma L'_1 - \nu'\cdot\alpha^t\sigma L_1 + \beta^t\sigma L'_0 - \beta'^t\sigma L_0 + \varkappa\beta'^t\sigma - \varkappa'\beta^t\sigma\big)\\ &+\big(-\nu L'_1\alpha + \nu' L_1\alpha - L'_0\beta + L_0\beta' + \varkappa\beta' - \varkappa'\beta\big)\alpha^t\sigma.\end{aligned}$$

Denote the second bracket by β''. Then the relations $L_1^t = -\sigma L_1\sigma^{-1}$, $L_0^t = -\sigma L_0\sigma^{-1}$ (obeyed by symplectic matrices) imply that the first bracket is equal to $\beta''^t\sigma$. Hence

$$L''_{-1} = (\alpha\beta''^t + \beta''\alpha^t)\sigma,$$

where

$$\beta'' = -\nu L'_1\alpha + \nu' L_1\alpha - L'_0\beta + L_0\beta' + \varkappa\beta' - \varkappa'\beta.$$

Prove that $\beta''^t\sigma\alpha = 0$. Making use of the above expression for $\beta''^t\sigma$ we find

$$\beta''^t\sigma\alpha = \nu\cdot\alpha^t\sigma L'_1\alpha - \nu'\cdot\alpha^t\sigma L_1\alpha + \beta^t\sigma L'_0\alpha - \beta'^t\sigma L_0\alpha + \varkappa\beta'^t\sigma\alpha - \varkappa'\beta^t\sigma\alpha.$$

The first two terms of the expression just obtained vanish by the relation (4.1.14) applied to L and L'. To the second pair of terms we apply the relations $L_0\alpha = \varkappa\alpha$ and $L'_0\alpha = \varkappa'\alpha$, after which all remaining terms vanish by the relations (4.1.12).

(d) Prove the relation (4.1.13) on the eigenvalues of the degree zero term. By definition,

$$(LL')_0 = \nu\alpha\alpha^t\sigma L'_2 + (\alpha\beta^t + \beta\alpha^t)\sigma L'_1 + L_0L'_0 + L_1(\alpha\beta'^t + \beta'\alpha^t)\sigma + \nu' L_2\alpha\alpha^t\sigma.$$

The last pair of terms obviously vanishes, having been multiplied by α from the right. We obtain

$$(LL')_0\alpha = \nu\alpha\alpha^t\sigma L'_2\alpha + \alpha\beta^t\sigma L'_1\alpha + \beta\alpha^t\sigma L'_1\alpha + \varkappa\varkappa'\alpha.$$

The third summand is equal to 0 by (4.1.14). Thus α is the eigenvector for a zero degree term even in the product LL':

$$(LL')_0\alpha = \alpha(\nu\cdot\alpha^t\sigma L'_2\alpha + \beta^t\sigma L'_1\alpha + \varkappa\varkappa'),$$

and in the commutator we obtain

$$L''_0\alpha = \alpha(\nu \cdot \alpha^t\sigma L'_2\alpha - \nu' \cdot \alpha^t\sigma L_2\alpha + \beta^t\sigma L'_1\alpha - \beta'^t\sigma L_1\alpha). \qquad (4.1.15)$$

(e) Conservation of the relation $\alpha^t\sigma L_1\alpha = 0$ in the product and commutator. For the product, by definition,

$$(LL')_1 = L_{-2}L'_3 + L_{-1}L'_2 + L_0L'_1 + L_1L'_0 + L_2L'_{-1} + L_3L'_{-2}.$$

Replacing L_{-2}, L'_{-2}, L_{-1}, L'_{-1} here by the known expressions we obtain

$$\begin{aligned}\alpha^t\sigma(LL')_1\alpha = {} & \nu(\alpha^t\sigma\alpha)\alpha^t\sigma L'_3\alpha + \big((\alpha^t\sigma\alpha)\beta^t + (\alpha^t\sigma\beta)\alpha^t\big)\sigma L'_2\alpha \\ & - \varkappa(\alpha^t\sigma L'_1\alpha) + \varkappa'(\alpha^t\sigma L_1\alpha) \\ & + \alpha^t\sigma L_2\big(\alpha(\beta'^t\sigma\alpha) + \beta'(\alpha^t\sigma\alpha)\big) + \nu'\alpha^t\sigma L_3\alpha(\alpha^t\sigma\alpha).\end{aligned}$$

By the relations (4.1.11)–(4.1.14) this expression vanishes. □

4.2 Almost-graded structure

In this section we consider the following cases: $\mathfrak{g} = \mathfrak{sl}(n)$, $\mathfrak{g} = \mathfrak{so}(n)$, $\mathfrak{g} = \mathfrak{sp}(2n)$, and, to conclude with, $\mathfrak{g} = \mathfrak{gl}(n)$ which is a somewhat special case with respect to the almost-gradeness. We assume for brevity that there are two P-points $P_\pm$. A general definition of the almost-graded structure is given in Section 1.4. A Lax operator algebra will be denoted by $\mathcal{L}$. It will also be denoted by $\overline{\mathfrak{g}}$ if we want to stress that it corresponds to a certain finite-dimensional Lie algebra $\mathfrak{g}$. For instance, we will use the notation $\overline{\mathfrak{gl}(n)}$ for the Lax operator algebra corresponding to $\mathfrak{g} = \mathfrak{gl}(n)$.

For all the algebras $\mathfrak{g}$, listed above, except for $\mathfrak{gl}(n)$, and $m \in \mathbb{Z}$ let

$$\mathcal{L}_m = \{L \in \mathcal{L} \mid (L) + D \geq 0\},$$

where (L) is the divisor of a $\mathfrak{g}$-valued function L, for $\mathfrak{g} = \mathfrak{sl}(n)$, $\mathfrak{g} = \mathfrak{so}(n)$

$$D = -mP_+ + (m+g)P_- + \sum_{s=1}^{ng} \gamma_s,$$

and for $\mathfrak{g} = \mathfrak{sp}(2n)$

$$D = -mP_+ + (m+g)P_- + 2\sum_{s=1}^{ng} \gamma_s.$$

To be more specific for (L), we should note that by order of a meromorphic matrix-valued function at a point we mean the minimal order of its entries.

We call $\mathcal{L}_m$ *the (homogeneous) subspace of degree m* of the Lie algebra $\mathcal{L}$. We will also denote the $\mathcal{L}_m$ by $\mathfrak{g}_m$ if $\mathcal{L} = \overline{\mathfrak{g}}$. For instance, $\overline{\mathfrak{so}(n)}_m$ is the degree m homogeneous subspace for $\mathcal{L} = \overline{\mathfrak{so}(n)}$.

Theorem 4.4. *For* $\mathfrak{g} = \mathfrak{sl}(n),\ \mathfrak{so}(n),\ \mathfrak{sp}(2n)$, *and* $\mathcal{L} = \overline{\mathfrak{g}}$

1°. $\dim \mathcal{L}_m = \dim \mathfrak{g}$;

2°. $\mathcal{L} = \bigoplus\limits_{m=-\infty}^{\infty} \mathcal{L}_m$;

3°. $[\mathcal{L}_k, \mathcal{L}_l] \subseteq \bigoplus\limits_{m=k+l}^{k+l+g} \mathcal{L}_m$.

Proof. First prove 1°. For $\mathfrak{g} = \mathfrak{sl}(n), \mathfrak{so}(n)$, the Riemann–Roch dimension of all $\mathfrak{g}$-valued functions L satisfying the relation $(L) + D \geq 0$ is equal to $(\dim \mathfrak{g})(ng + 1)$. For $\mathfrak{g} = \mathfrak{sp}(2n)$ it is equal to $(\dim \mathfrak{g})(2ng + 1)$. We will prove that for $\mathfrak{g} = \mathfrak{sl}(n)$, $\mathfrak{g} = \mathfrak{so}(n)$, and an arbitrary $m \in \mathbb{Z}$ there are exactly $\dim \mathfrak{g}$ relations at every pole γ_s, and for $\mathfrak{g} = \mathfrak{sp}(2n)$ the number of those relations is $2 \dim \mathfrak{g}$. This means that $\dim \mathcal{L}_m = \dim \mathfrak{g}$.

Consider first the case $\mathfrak{g} = \mathfrak{sl}(n)$. The elements of the subspace $\mathcal{L}_m$ satisfy certain conditions of three kinds coming from (4.1.2), (4.1.3): on the residues, eigenvalues, and traces of the matrix $L \in \mathcal{L}_m$, as follows:

(1) At every γ-point, $L_{-1} = \alpha\beta^t$ which would give $\dim \mathfrak{g}$ relations (since $L_{-1} \in \mathfrak{g}$) if the right-hand side is fixed. However, the latter depends on the free n-dimensional vector β. Hence we have $\dim \mathfrak{g} - n$ conditions at every one of the ng simple poles γ_s.

(2) At every γ-point, $L_0\alpha = \varkappa\alpha$ which gives n conditions. Taking account of one free parameter $\varkappa$ we obtain $n - 1$ condition at every γ_s.

(3) We also have $\operatorname{tr} L = 0$, i.e., one more relation at every weak singularity.

Thus we have $(\dim \mathfrak{g} - n) + (n - 1) + 1 = \dim \mathfrak{g}$ relations at every γ_s, as required.

For $\mathfrak{g} = \mathfrak{so}(n)$ we follow the same line of arguing. Again, the relation $L_{-1} = \alpha\beta^t - \beta\alpha^t$ (4.1.6) gives $\dim \mathfrak{g} - n$ equations, the relation $L_0\alpha = \varkappa\alpha$ (4.1.8) gives $n-1$ equation, and $\beta^t\alpha = 0$ (4.1.7) gives one more equation. Summing up, we obtain $\dim \mathfrak{g}$ equations at every point γ_s.

In the case $\mathfrak{g} = \mathfrak{sp}(2n)$, at every point γ_s we have the following conditions:

$L_{-2} = \nu\alpha\alpha^t$:	$\dim \mathfrak{g} - 1$	conditions (one free parameter ν)
$L_{-1} = (\alpha\beta^t + \beta\alpha^t)\sigma$:	$\dim \mathfrak{g} - 2n$	conditions ($2n$ free parameters β)
$L_0\alpha = \varkappa\alpha$:	$2n - 1$	conditions (one free parameter $\varkappa$)
$\beta^t\sigma\alpha = 0,\ \ \alpha^t\sigma L_1\alpha = 0$:	2	conditions,

i.e., $2 \dim \mathfrak{g}$ conditions at every one of the ng points, as required.

For $m > 0$ and $m < -g$ the subspaces $\mathcal{L}_m$ are linear-independent for the obvious reason that the orders at $P_\pm$ of the elements of those subspaces are different for different m.

For $-g \le m \le 0$ the linear independence of $\mathcal{L}_m$ follows from the fact that there are no Lax operators which are *holomorphic everywhere* . We would like to emphasize that the last argument applies only to the case of simple Lie algebras. This explains why the case of the reductive Lie algebra $\mathfrak{gl}(n)$ requires some modification (see below).

The statement 3° of the theorem follows from the consideration of orders at the points $P_\pm$. □

Theorem 4.4 defines an *almost-graded structure* on $\mathcal{L}$.

Consider now the case $\mathfrak{g} = \mathfrak{gl}(n)$. In this case $\mathcal{L}$ contains the subspace of functions taking values in the 1-dimensional space of scalar matrices. Let L be such a function. By (4.1.2) we obtain $\operatorname{tr} L_{-1} = 0$. Since L_{-1} is scalar, we obtain $L_{-1} = 0$. Hence, L is holomorphic except at $P_\pm$. Let $\mathcal{A}$ denote the algebra of meromorphic functions on Σ holomorphic except at $P_\pm$. Then $L \in \mathcal{A} \cdot id$ where id is the unit matrix. Therefore,

$$\overline{\mathfrak{gl}(n)} = \overline{\mathfrak{sl}}(n) \oplus \mathcal{A} \cdot id. \tag{4.2.1}$$

In Section 1.3 the Krichever-Novikov base $\{A_m\}$ in the space of such functions was introduced. Denote $\mathcal{A}_m = \mathbb{C} A_m$ and set $\mathfrak{gl}(n)_m = \mathfrak{sl}(n)_m \oplus (\mathcal{A}_m \cdot id)$. For $m > 0$ and $m < -g$ this definition is equivalent to the definition of $\mathcal{L}_m$ (with $\mathfrak{g} = \mathfrak{gl}(n)$) given above.

With $\mathcal{L}_m = \mathfrak{gl}(n)_m$, Theorem 4.4 remains true for $\mathfrak{g} = \mathfrak{gl}(n)$, as it follows from (4.2.1). The relation 3° of the theorem holds with the different upper limit of summation determined by the algebra $\mathcal{A}$ (see [26]).

4.3 Central extensions of Lax operator algebras: the construction

The conventional expression for the 2-cocycle giving the almost-graded central extension for an affine Krichever–Novikov current algebra (in particular, a loop algebra) is $\operatorname{tr}\operatorname{res}_{P_+} L\, dL'$. For those algebras the cocycle is local. In the case of Lax operator algebras, the cocycle above is by no means local. In this section we improve it in order to obtain a local cocycle.

For every operator L the eigenvalue $\varkappa$ of its zero-degree component L_0 (see (4.1.3)) can be considered as a linear functional of L. We denote this functional by $\varkappa(L)$.

Lemma 4.5. *At every point γ of the Tyurin divisor, the one-form* $\operatorname{tr} L\,dL'$ *has at most a simple pole, and*

$$\operatorname{res}_{z=0} \operatorname{tr} L\,dL' = \begin{cases} \varkappa([L,L']) & \text{if} \quad \mathfrak{g} = \mathfrak{gl}(n), \\ 2\varkappa([L,L']) & \text{if} \quad \mathfrak{g} = \mathfrak{so}(n)(n) \quad \text{or} \quad \mathfrak{g} = \mathfrak{sp}(2n). \end{cases} \tag{4.3.1}$$

Proof. We prove the statement case by case, by means of a straightforward calculation of both parts of the relation (4.3.1).

First, consider the case $\mathfrak{g} = \mathfrak{gl}(n)$.

Using

$$dL' = \left(-\frac{\alpha\beta'^t}{z^2} + L_1' + \cdots\right) dz,$$

and (4.1.1) we obtain

$$L\,dL' = \left(-\frac{\alpha\beta^t\alpha\beta'^t}{z^3} - \frac{L_0\alpha\beta'^t}{z^2} - \frac{L_1\alpha\beta'^t - \alpha\beta^t L_1'}{z} + \cdots\right) dz.$$

The first term vanishes since $\beta^t\alpha = 0$. The second term vanishes under the symbol of trace, since $L_0\alpha = \varkappa\alpha$, and $\operatorname{tr}\alpha\beta'^t = \beta'^t\alpha = 0$. The third term gives us the required residue. We have

$$\operatorname{res}_{z=0} \operatorname{tr} L\,dL' = \operatorname{tr}(\alpha\beta^t L_1' - L_1\alpha\beta'^t) = \beta^t L_1'\alpha - \beta'^t L_1\alpha.$$

Now calculate the right-hand side of (4.3.1). Denote by $[L,L']_0$ the zero-degree coefficient of the expansion (4.1.1) for the commutator $[L,L']$. We have

$$[L,L']_0 = \alpha\beta^t L_1' + L_0L_0' + L_1\alpha\beta'^t - \alpha\beta'^t L_1 - L_0'L_0 - L_1'\alpha\beta^t.$$

Multiply both parts of this relation by α from the right-hand side. Then the third terms in every row vanish, since they contain the factors $\beta'^t\alpha$, $\beta^t\alpha$, which are zeroes by (4.1.2). The second terms annihilate since $L_0L_0'\alpha = \varkappa'\varkappa\alpha$ and $L_0'L_0\alpha = \varkappa\varkappa'\alpha$. Hence

$$[L,L']_0\alpha = \alpha\beta^t L_1'\alpha - \alpha\beta'^t L_1\alpha = \alpha(\beta^t L_1'\alpha - \beta'^t L_1\alpha).$$

The expression in brackets is a 1×1-matrix, i.e., a complex number. Certainly, this complex number is the eigenvalue of $[L,L']_0$ on the vector α, i.e., $\varkappa([L,L'])$. Its value exactly meets the resulting expression above for $\operatorname{tr}\operatorname{res}_{z=0} L\,dL'$, which completes the proof.

Now consider the case $\mathfrak{g} = \mathfrak{so}(n)$. Using (4.1.5) and the relation

$$dL' = (-L_{-1}'z^{-2} + L_1' + \cdots)\,dz,$$

where L_{-1}' is given by (4.1.6), we obtain

$$L\,dL' = \left(-\frac{L_{-1}L_{-1}'}{z^3} - \frac{L_0L_{-1}'}{z^2} - \frac{L_1L_{-1}' - L_{-1}L_1'}{z} + \cdots\right) dz. \tag{4.3.2}$$

For the first term we have

$$L_{-1}L'_{-1} = (\alpha\beta^t - \beta\alpha^t)(\alpha\beta''^t - \beta'\alpha^t) = \alpha(\beta^t\alpha)\beta''^t - \beta(\alpha^t\alpha)\beta''^t - \alpha\beta^t\beta'\alpha^t - \beta\alpha^t\beta'\alpha^t.$$

The first two summands vanish due to (4.1.7). For the remainder, we have

$$\mathrm{tr}\,(L_{-1}L'_{-1}) = \mathrm{tr}\,(-\alpha\beta^t\beta'\alpha^t - \beta\alpha^t\beta'\alpha^t) = -(\alpha^t\alpha)\beta^t\beta' - (\alpha^t\beta)\alpha^t\beta'$$

which again vanishes due to (4.1.7).

Again, the term in (4.3.2) containing z^{-2} vanishes under the symbol of trace. Indeed, by definition,

$$L_0L'_{-1} = L_0(\alpha\beta''^t - \beta'\alpha^t) = \varkappa\alpha\beta''^t - L_0\beta'\alpha^t.$$

Now observe that $\mathrm{tr}\,(L_0\beta'\alpha^t) = \mathrm{tr}\,(\alpha^t L_0\beta')$ and $\alpha^t L_0 = -\varkappa\alpha^t$. Hence, $\mathrm{tr}\,(L_0L'_{-1}) = 2\varkappa\alpha^t\beta'$ which vanishes due to (4.1.7).

The third term in (4.3.2) gives us the required residue. We have

$$\mathrm{res}_{z=0}\, L\,dL' = (L_{-1}L'_1 - L_1L'_{-1}).$$

Substituting L_{-1}, L'_{-1} from (4.1.6), we have

$$\mathrm{res}_{z=0}\, L\,dL' = \alpha\beta^t L'_1 - \beta\alpha^t L'_1 - L_1\alpha\beta''^t + L_1\beta'\alpha^t,$$

and hence

$$\mathrm{tr}\ \mathrm{res}_{z=0}\, L\,dL' = \beta^t L'_1\alpha - \alpha^t L'_1\beta - \beta''^t L_1\alpha + \alpha^t L_1\beta'.$$

Due to the skew-symmetry of matrices L_1, L'_1 the first two summands of the last relation are equal, and the same is true for the last two summands. Hence

$$\mathrm{tr}\ \mathrm{res}_{z=0}\, L\,dL' = 2(\beta^t L'_1\alpha - \beta''^t L_1\alpha).$$

From (4.1.9) obviously follows that $[L, L']_0\alpha = \beta^t L'_1\alpha - \beta''^t L_1\alpha$ which proves the lemma.

Finally, consider the case $\mathfrak{g} = \mathfrak{sp}(2n)$.

A straightforward calculation of LdL' based on the expansion (4.1.10) shows that the terms with z^{-5} and z^{-4} of that matrix-valued 1-form are equal to 0 by the relations (4.1.12).

For the term with z^{-3} we have

$$(LdL')_{-3} = -(\beta^t\sigma\beta' - 2\nu')\alpha\alpha^t\sigma.$$

This expression vanishes under the symbol of trace:

$$\mathrm{tr}\,(LdL')_{-3} = -(\beta^t\sigma\beta' - 2\nu')(\alpha^t\sigma\alpha) = 0.$$

Similarly, using (4.1.12)-(4.1.14) we have

$$\mathrm{tr}\,(LdL')_{-2} = \nu(\alpha^t\sigma L'_1\alpha) - 2\varkappa(\beta''^t\sigma\alpha) - \nu'(\alpha^t\sigma L_1\alpha) = 0.$$

Thus, the 1-form $\operatorname{tr} LdL'$ indeed has at most a simple pole at the point in question. A straightforward calculation results in

$$\operatorname{tr}(LdL')_{-1} = 2(\nu \cdot \alpha^t \sigma L_2' \alpha - \nu' \cdot \alpha^t \sigma L_2 \alpha + \beta^t \sigma L_1' \alpha - \beta'^t \sigma L_1 \alpha),$$

which exactly coincides with the double expression (4.1.15) for $\varkappa([L, L'])$. □

We want to eliminate the singularities of the 1-form $\operatorname{tr} LdL'$ at γ_s by subtracting another expression for the eigenvalue on the right-hand side of (4.3.1). Remarkably, we are able to give this new expression in terms of connections in the holomorphic vector bundles on Σ, given in explicit form in [32]. In the following, ω denotes the 1-form of such a connection.

Lemma 4.6. *Let ω be a $\mathfrak{g}$-valued 1-form such that locally in the neighborhood of a point γ*

$$\omega = \omega_{-1}\frac{dz}{z} + \omega_0 dz + \cdots,$$

and ω satisfies the similar requirements as L, namely

$$\omega_{-1} = \begin{cases} \alpha\tilde{\beta}^t & \text{if } \mathfrak{g} = \mathfrak{gl}(n), \\ \alpha\tilde{\beta}^t - \tilde{\beta}\alpha^t, \quad \alpha^t\alpha = 0 & \text{if } \mathfrak{g} = \mathfrak{so}(n), \\ (\alpha\tilde{\beta}^t + \tilde{\beta}\alpha^t)\sigma, \quad \alpha^t\sigma\omega_1\alpha = 0 & \text{if } \mathfrak{g} = \mathfrak{sp}(2n), \end{cases}$$

and $\omega_0\alpha = \tilde{\varkappa}\alpha$, with the only one modification: $\tilde{\beta}^t\alpha = 1$ ($\tilde{\beta}^t\sigma\alpha = 1$ for $\mathfrak{g} = \mathfrak{sp}(2n)$). Then the 1-form $\operatorname{tr} L\omega$ *has at most a simple pole at $z = 0$, and*

$$\operatorname{res}_{z=0} \operatorname{tr} L\omega = \varkappa(L).$$

Proof. For $\mathfrak{g} = \mathfrak{gl}(n)$ we have

$$L = \frac{\alpha\beta^t}{z} + L_0 + \cdots, \quad \omega = \left(\frac{\alpha\tilde{\beta}^t}{z} + \omega_0 + \cdots\right)dz,$$

hence

$$L\omega = \left(\frac{\alpha\beta^t\alpha\tilde{\beta}^t}{z^2} + \frac{\alpha\beta^t\omega_0 + L_0\alpha\tilde{\beta}^t}{z} + \cdots\right)dz.$$

As above, the first term vanishes. For the second term we have

$$\operatorname{res}_{z=0} tr\, L\omega = \operatorname{tr}(\alpha\beta^t\omega_0 + L_0\alpha\tilde{\beta}^t) = \beta^t\omega_0\alpha + \tilde{\beta}^t L_0\alpha.$$

Then, $\omega_0\alpha = \tilde{\varkappa}\alpha$, $L_0\alpha = \varkappa\alpha$, hence $\beta^t\omega_0\alpha = \tilde{\varkappa}\beta^t\alpha = 0$ and $\tilde{\beta}^t L_0\alpha = \varkappa\tilde{\beta}^t\alpha = \varkappa$. This completes the proof in the case $\mathfrak{g} = \mathfrak{gl}(n)$.

In the case $\mathfrak{g} = \mathfrak{so}(n)$ the coefficient at the degree -2 term in the product $L\omega$ is equal, by definition, to

$$(\alpha\beta^t - \beta\alpha^t)(\alpha\tilde{\beta}^t - \tilde{\beta}\alpha^t) = \alpha(\beta^t\alpha)\tilde{\beta}^t - \alpha\beta^t\tilde{\beta}\alpha^t - \beta(\alpha^t\alpha)\tilde{\beta}^t + \beta\alpha^t\tilde{\beta}\alpha^t.$$

The first and the third terms vanish due to (4.1.7). For the trace of the remainder we have

$$\operatorname{tr}(-\alpha\beta^t\tilde{\beta}\alpha^t + \beta\alpha^t\tilde{\beta}\alpha^t) = -(\alpha^t\alpha)\beta^t\tilde{\beta} + (\alpha^t\beta)\alpha^t\tilde{\beta}$$

which vanishes for the same reason.

Multiplying expansions for L and ω we find

$$\operatorname{tr}_{z=0}\operatorname{res}(L\omega) = \operatorname{tr}(L_{-1}\omega_0 + L_0\omega_{-1}) = \operatorname{tr}(\alpha\beta^t - \beta\alpha^t)\omega_0 + \operatorname{tr} L_0(\alpha\tilde{\beta}^t - \tilde{\beta}\alpha^t).$$

For the first summand we have

$$\operatorname{tr}(\alpha\beta^t - \beta\alpha^t)\omega_0 = \operatorname{tr}(\alpha\beta^t\omega_0 - \beta\alpha^t\omega_0) = \beta^t\omega_0\alpha - \alpha^t\omega_0\beta.$$

Observe that, by skew-symmetry, $\alpha^t\omega_0\beta = -\beta^t\omega_0\alpha$; hence

$$\operatorname{tr}(\alpha\beta^t - \beta\alpha^t)\omega_0 = \beta^t\omega_0\alpha - \alpha^t\omega_0\beta = 2\beta^t\omega_0\alpha = 2\tilde{\varkappa}\beta^t\alpha,$$

which vanishes due to (4.1.7).

For the second summand we have

$$\operatorname{tr} L_0(\alpha\tilde{\beta}^t - \tilde{\beta}\alpha^t) = \operatorname{tr}(L_0\alpha\tilde{\beta}^t - L_0\tilde{\beta}\alpha^t) = \tilde{\beta}^t L_0\alpha - \alpha^t L_0\tilde{\beta}.$$

Since $L_0\alpha = \varkappa\alpha$, $\alpha^t L_0 = -\varkappa\alpha^t$ and $\tilde{\beta}^t\alpha = \alpha^t\tilde{\beta} = 1$ we obtain

$$\operatorname{tr} L_0(\alpha\tilde{\beta}^t - \tilde{\beta}\alpha^t) = 2\varkappa.$$

Finally, consider the case $\mathfrak{g} = \mathfrak{sp}(2n)$. The expansion for $L\omega$ begins with z^{-3}. We have

$$\begin{aligned}
&\operatorname{tr}(L\omega)_{-3} = \nu(\tilde{\beta}^t\sigma\alpha)(\alpha^t\sigma\alpha) = 0,\\
&\operatorname{tr}(L\omega)_{-2} = \nu(\alpha^t\sigma\tilde{\beta})(\alpha^t\sigma\alpha) + (\beta^t\sigma\tilde{\beta})(\alpha^t\sigma\alpha) + (\alpha^t\sigma\tilde{\beta})(\alpha^t\sigma\beta) = 0.
\end{aligned}$$

Thus, the 1-form $\operatorname{tr} L\omega$ indeed has at most a simple pole at the point in question. The calculation of the residue results in

$$\operatorname{tr}(L\omega)_{-1} = \nu\cdot\alpha^t\sigma\omega_1\alpha + \beta^t\sigma(\omega_0\alpha) + (\alpha^t\sigma\omega_0)\beta + \tilde{\beta}^t\sigma(L_0\alpha) + (\alpha^t\sigma L_0)\tilde{\beta}.$$

By assumption of the lemma, $\alpha^t\sigma\omega_1\alpha = 0$, $\omega_0\alpha = \tilde{\varkappa}\alpha$. This also implies that $\alpha^t\sigma\omega_0 = -\tilde{\varkappa}\alpha^t\sigma$. For this reason,

$$\operatorname{tr}(L\omega)_{-1} = \tilde{\beta}^t\sigma(L_0\alpha) + (\alpha^t\sigma L_0)\tilde{\beta}.$$

By the relations $L_0\alpha = \varkappa\alpha$, $\alpha^t\sigma L_0 = -\varkappa\alpha^t\sigma$ we have

$$\operatorname{tr}(L\omega)_{-1} = 2\varkappa(\tilde{\beta}^t\sigma\alpha) = 2\varkappa. \qquad \square$$

Theorem 4.7. *For every ω satisfying the conditions of Lemma 4.6, the 1-form $\operatorname{tr}(L\,dL' - [L, L']\omega)$ is regular except at the points $P_1, \dots, P_N$, and the expression*

$$\gamma(L, L') = \operatorname{res}_{P_+} \operatorname{tr}(L\,dL' - [L, L']\omega)$$

gives a local cocycle on the Lax operator algebra.

Proof. In course of proofs of Lemma 4.5 and Lemma 4.6, we have seen that the 1-forms $\operatorname{tr} L\,dL'$ and $\operatorname{tr}[L, L']\omega$ have simple poles at every point γ_s, and that their residues are equal to the same quantity $\varkappa_s([L, L'])$. Hence, their difference is regular at every γ_s.

Let us assume for brevity that there are only two P-points, which are denoted by $P_\pm$. Assume that at point P_+ we have

$$L(z) = \sum_{i=m}^{\infty} a_i z^i,\ L'(z) = \sum_{j=m'}^{\infty} b_j z^j,\ \omega(z) = \sum_{k=m_+}^{\infty} c_k z^k dz. \tag{4.3.3}$$

Then

$$L(z)\,dL'(z) = \sum_{p=m+m'}^{\infty} \Big(\sum_{i+j=p} j a_i b_j \Big) z^{p-1} dz$$

and

$$[L(z), L'(z)]\omega = \sum_{p=m+m'+m_+}^{\infty} \Big(\sum_{i+j+k=p} [a_i, b_j] c_k \Big) z^p dz.$$

In order for one of these 1-forms to have a nontrivial residue at the point P_+, it is necessary that either $m + m' \le 0$ or $m + m' + m_+ \le -1$; in other words:

$$m + m' \le \max\{0, -1 - m_+\}.$$

If L and L' are homogeneous of degrees m, m', then at the point P_- their expansions (similar to (5.1.3)) begin with $i = -m - g$, $j = -m' - g$, respectively. The expansion for ω begins with some integer m_-. Hence the condition at P_- appears as

$$-m - m' - 2g \le \max\{0, -1 + m_-\}.$$

Finally, we obtain

$$\min\{0, 1 - m_-\} - 2g \le m + m' \le \max\{0, -1 - m_+\}.$$

Since $m_\pm$ are fixed (ω is fixed), this means exactly that the cocycle is local. $\square$

4.4 Uniqueness theorem

The uniqueness results for the central extensions of Lax operator algebras are quite similar to those for affine Krichever-Novikov algebras (Sections 1.6 and 1.8). They literally become the same if we replace the differentials with covariant derivatives with respect to the connection form ω (introduced in Lemma 4.6) in all expressions for the geometrical cocycles in Sections 1.6 and 1.8 (see the relations (4.4.2), (4.4.3) below). Here we make a few remarks on the proof of the uniqueness results. For more detail see [54], as well as the forthcoming book [50].

Recall that the Krichever–Novikov current algebras are of the form $\mathfrak{g}\otimes\mathcal{A}$ where $\mathcal{A}$ is an associative commutative algebra, and $\mathfrak{g}$ is a simple Lie algebra. We will refer to this fact as splitting to the tensor product. For splitting Lie algebras Ch. Kassel [18] showed that their universal central extensions are parameterized by Kähler differentials modulo exact differentials. In particular, it is not necessarily one-dimensional (this was explained in another way in Chapter 1). Therefore, in general one-dimensional central extensions are not unique. The uniqueness is restored if one considers central extensions in the category of almost-graded Lie algebras. The uniqueness results for the affine Krichever–Novikov algebras hardly use the result due to Kassel [48, 49].

Unlike Krichever–Novikov algebras, the Lax operator algebras do not split to tensor products. The result due to Kassel is no longer applicable. For this reason, new approaches have been elaborated in [54] to prove the same uniqueness results. Lax operator algebras provide a most general known point of view on the setting the problem of classification of central extensions of 1-dimensional current algebras, and methods of its solution. Here we only consider the two-point case, i.e., $D = P_+ + P_-$. The principal structure of the multipoint case is similar and, I believe, will be considered in [50].

Let us briefly describe the uniqueness results. Let $\mathcal{V}$ be the Krichever–Novikov vector field algebra, as above. Making use of the connection $\nabla^{(\omega)}$, we introduce the $\mathcal{V}$-action on the Lax operator algebra $\mathfrak{L}$: we set $\nabla^{(\omega)}L = dL + [\omega, L]$, $L \in \mathfrak{L}$, and $\nabla_e^{(\omega)}L = eL + [e\omega, L]$, $e \in \mathcal{V}$. The Lax operator algebra turns out to be an almost-graded $\mathcal{V}$-module with respect to this action, and the differential operators $\nabla_e^{(\omega)}$ are holomorphic outside $\{P_+, P_-\}$.

The $\mathcal{V}$-module structure given by the connection $\nabla^{(\omega)}$ enables us to generalize the important notion of $\mathcal{V}$-invariant cocycles introduced above. A cocycle γ on $\mathfrak{L}$ is called $\mathcal{V}$-invariant if

$$\gamma(\nabla_e^{(\omega)}L, L') + \gamma(L, \nabla_e^{(\omega)}L') = 0 \tag{4.4.1}$$

for all $e \in \mathcal{V}$, $L, L' \in \mathfrak{L}$.

It can be easily shown that in terms of $\nabla^{(\omega)}$ the cocycles introduced in Section 4.3 (Theorem 4.7) read as

$$\gamma_{1,\omega,C}(L, L') = \frac{1}{2\pi \mathrm{i}} \int_C \operatorname{tr}(L \cdot \nabla^{(\omega)} L'). \tag{4.4.2}$$

The corresponding analog of the cocycle γ_2 in (1.8.3) is then written as

$$\gamma_{2,\omega,C}(L, L') = \frac{1}{2\pi \mathrm{i}} \int_C \operatorname{tr}(L) \cdot \operatorname{tr}(\nabla^{(\omega)} L'). \tag{4.4.3}$$

We will continue referring to them as to *geometric cocycles*. Here C is a separating cycle on Σ, avoiding the points of the Tyurin divisor. It was shown in [54] that the corresponding cohomology classes do not depend on the choice of the connection. Besides the fact that the cocycles (4.4.2) and (4.4.3) are local, it turns out that they are $\mathcal{V}$-invariant.

We call a 2-cohomology class *$\mathcal{V}$-invariant* if it contains a $\mathcal{V}$-invariant cocycle, and *local* if it contains a local cocycle. We stress that other representatives of a local ($\mathcal{V}$-invariant) cohomology class by no means possess the corresponding property. In the case of a simple Lie algebra $\mathfrak{g}$, a cohomology class contains at most one $\mathcal{V}$-invariant cocycle. As above, here we are only interested in almost-graded central extensions. Recall from Chapter 1 that the almost-graded central extensions are in one-to-one correspondence with the local $\mathcal{V}$-invariant cohomology classes. Thus the classification of the almost-graded central extensions is given by the following theorem.

Theorem 4.8.

(a) *If $\mathfrak{g}$ is simple (i.e., $\mathfrak{g} = \mathfrak{sl}(n), \mathfrak{so}(n), \mathfrak{sp}(2n)$), then the space of local cohomology classes on $\mathcal{L} = \overline{\mathfrak{g}}$ is one-dimensional. If we fix any connection form ω, then the space will be generated by the class of $\gamma_{1,\omega}$. Every $\mathcal{V}$-invariant (with respect to the connection ω) local cocycle is a multiple of $\gamma_{1,\omega}$.*

(b) *For $\mathfrak{g} = \mathfrak{gl}(n)$ the space of local cohomology classes on $\mathcal{L} = \overline{\mathfrak{g}}$ which are $\mathcal{V}$-invariant, having been restricted to the scalar subalgebra, is two-dimensional. If we fix any connection form ω, then the space will be generated by the classes of cocycles $\gamma_{1,\omega}$ and $\gamma_{2,\omega}$. Every $\mathcal{V}$-invariant local cocycle is a linear combination of $\gamma_{1,\omega}$ and $\gamma_{2,\omega}$.*

The proof of the first part of the theorem (which treats the dimension of the local cohomology space) relies on the structure theory of simple Lie algebras over $\mathbb{C}$ and the properties of the Chevalley basis. We show [54] that, up to the coboundary, a local cocycle is determined by its value $\gamma(H_1^\alpha, \gamma(H_{-1}^\alpha)$, where $H^\alpha \in \mathfrak{g}$ is the Chevalley generator of the Cartan type, corresponding to a simple root α (which may be chosen arbitrarily, but is previously fixed), $H_n^\alpha \in \mathcal{L}$ is the unique element in the degree n homogeneous subspace such that $H_n^\alpha(z_+) = H^\alpha z_+^n + O(z_+^{n+1})$ (z_+ is a local coordinate in the neighborhood of P_+). Hence the local cohomology space is at most 1-dimensional. As stated above, the cocycle $\gamma_{1,\omega}$ (4.4.2) is local. Hence the class of $\gamma_{1,\omega}$ generates the local cohomology space.

For the loop algebra $\mathfrak{g} \otimes \mathbb{C}[z, z^{-1}]$, as a very special case of Lax operator algebra, the thread of our arguments becomes simpler and is similar to the arguments of Garland [9].

To prove the uniqueness up to a scalar factor of the $\mathcal{V}$-invariant and local cocycle in Theorem 4.8(a) is a different story altogther. The proof [54] follows the general strategy developed in [48]. By a certain recurrent procedure using locality and $\mathcal{V}$-invariance, we show that the space of the $\mathcal{V}$-invariant local cocycles has the same dimension as the space of the symmetric invariant bilinear forms on $\mathfrak{g}$. In the case where $\mathfrak{g}$ is simple, the uniqueness of the cocycle is immediately implied. On the other hand, for $\mathfrak{g} = \mathfrak{gl}(n)$ there are two linear-independent invariant forms which result in the appearance of $\gamma_{2,\omega}$ (Theorem 4.8(b)).

Chapter 5

Lax equations with a spectral parameter on Riemann surfaces, and their hierarchies

Similar to the characterization of Krichever–Novikov algebras as algebras of gauge and conformal symmetries of the 2D CFT, we may characterize Lax operator algebras as the algebras of the integrable systems theory. This chapter deals with the construction of a broad class of finite-dimensional integrable systems, their hierarchies of commuting flows, and their Hamiltonian theory, in terms of the Lax operator algebras. It can serve as an introduction to the theory of such integrable systems, including Hitchin systems, gyroscopes, and integrable cases of flow around a solid.

The main, and quite intriguing, conclusion of this chapter is as follows. The relations implying closure of the Lax operator algebras with respect to their brackets (Section 4.1), i.e., (4.1.1)–(4.1.3) for $\mathfrak{g} = \mathfrak{gl}(n)$, (4.1.5)–(4.1.8) for $\mathfrak{g} = \mathfrak{so}(n)$, (4.1.10)–(4.1.14) for $\mathfrak{g} = \mathfrak{sp}(2n)$, are exactly what is needed for the development of the Hamiltonian theory for the corresponding integrable systems.

As already mentioned in the Preface, the key step was done by I. M. Krichever, who in [33] proposed the theory of Lax operators with the spectral parameter on a Riemann surface. In particular, it has resulted in the development of the theory of Lax operator algebras described in the previous chapter. From the point of view of Lax operator algebras what was done by I. M. Krichever corresponds to the case $\mathfrak{g} = \mathfrak{gl}(n)$. In [69–71] we have posed and resolved the problem of the generalization of the theory due to Krichever to all Lax operator algebras. The result of this generalization is presented here.

In the present chapter we focus at the dependence of Lax operators on Tyurin parameters and their dual quantities. To distinguish between this point of view and that adopted in the previous chapter, we introduce the term L-operator for a Lax operator corresponding to any fixed set of Tyurin parameters. Thus by Lax operator we mean the mapping which sends a point of the cotangent bundle on the space of Tyrin data (extended by marked points) to an element of the corresponding Lax operator algebra. With every positive divisor $D = \sum m_i P_i$ we associate the subspace $\mathcal{L}^D$ of the cotangent bundle, such that the Lax operator, as a function on Σ, has a pole of the order of at most m_i at P_i for every fixed point in that subspace. We consider a certain dynamical system on $\mathcal{L}^D$ given by the Lax equation. We prove that these dynamical systems are Hamiltonian and integrable with respect to the Krichever–Phong symplectic structure.

In Section 5.1, following [33, 70] we introduce *M-operators* (partners of Lax operators in Lax pairs) and study their analytic and algebraic properties. Every M-operator defines a flow in the cotangent bundle on the space of extended Tyurin data. The corresponding dynamics of the Tyurin parameters (5.1.7) plays a fundamental role. It appeared in [25] as an ansatz for effective integrating the Kadomtsev–Petviashvili equation in the class of high rank solutions, and is hardly used in [33].

In Section 5.2 we give an alternative definition of L-operators. Following [33, 70] we introduce them as M-operators yielding a trivial dynamics of Tyurin data.

In Section 5.3 we prove that a Lax equation is equal to the system of equations on the main parts of the Lax operator at the points P_i, and of the above mentioned dynamics equations of Tyurin data. We prove a criteria of well-definiteness of the dynamical system given by a Lax equation on a space $\mathcal{L}^D$.

In Section 5.4 we construct hierarchies of commuting flows on the cotangent bundle on $\mathcal{L}^D$ given by a Lax operator. A key role is played by Lemma 5.5, which treats the dimension of the space of M-operators which satisfy the criteria of Section 5.3. Modulo that lemma, we follow the lines of [33]. We prove the final result for all classical Lie algebras in question, in particular for $\mathfrak{sp}(2n)$, which was not done in the author's earlier article [70] on hierarchies. We would like to stress here that in general L and M do not belong to the same Lie algebra, we give the list of the corresponding pairs of Lie algebras (5.3.2). Observe that the method of constructing a hierarchy given in this section is not unique. In Section 5.7 we give another one for elliptic curves, enabling us to incorporate certain Calogero–Moser systems within the framework of our approach.

In Sections 5.5 and 5.6 we develop a Hamiltonian theory for the Lax equations in question. In Section 5.5 we construct the analog of the Krichever–Phong symplectic structure on the cotangent bundle of a certain subspace $\mathcal{P}^D \subset \mathcal{L}^D/G$ (where $\mathfrak{g} = Lie\,G$) invariant with respect to the flows of our commuting hierarchy. In Section 5.6 we prove that the hierarchies of Section 5.4 are Hamiltonian with respect to that structure, and we construct the corresponding Hamiltonians. Again, the presentation is similar for all classical Lie algebras in question and follows the lines of [33], except for the proofs of the nondegenerateness of the Krichever–Phong symplectic form (Section 5.5) and of the holomorphy of the spectra of operators in Lax pairs (Section 5.6). The latter is done individually for every class of Lie algebras. We also give the proofs of some statements in [33] in slightly more detail.

In Section 5.7 we consider four examples of integrable systems in the framework of our approach. The first one is the elliptic Calogero–Moser system for $\mathfrak{g} = \mathfrak{gl}(n)$. This was considered earlier in [24, 33]. Further on, by modifying the technique of Section 5.4 we obtain the Lax operators and the corresponding Hamiltonians hierarchies for the elliptic Calosero–Moser $\mathfrak{so}(2n)$, $\mathfrak{so}(2n+1)$, and $\mathfrak{sp}(2n)$ systems. Observe that in [13] Lax representations with a spectral parameter on an elliptic curve are obtained for all irreducible reduced root systems. They are of the Backer–Akhieser type, while our Lax operators are meromorphic. Nonetheless, they are gauge equivalent; see Section 5.7 for more details.

5.1 *M*-operators

Let Σ be a compact Riemann surface of genus g with two marked points P_+, P_-, and $\mathfrak{g}$ be a Lie algebra over $\mathbb{C}$ from the following: $\mathfrak{gl}(n)$, $\mathfrak{sl}(n)$, $\mathfrak{so}(2n)$, $\mathfrak{so}(2n+1)$, $\mathfrak{sp}(2n)$, $\mathfrak{s}(n)$, $\mathfrak{tsp}(2n)$, where $\mathfrak{s}(n)$ is the algebra of scalar matrices, and $\mathfrak{tsp}(2n)$ is a Lie subalgebra of $\mathfrak{sp}(2n+2)$ consisting of the matrices having a zero first column and last row.

Let us fix K additional points $\gamma_s \in \Sigma$, and let

$$W := \{\gamma_s \in \Sigma \setminus \{P_+, P_-\} \mid s = 1, \dots, K\} \tag{5.1.1}$$

(K will be specified in Section 5.4). To every point γ_s we assign a vector $\alpha_s \in \mathbb{C}^p$ given up to a scalar factor where p is the dimension of the standard (vector) representation of the corresponding $\mathfrak{g}$ (i.e., the elements of $\mathfrak{g}$ are $p\times p$-matrices). The system

$$T := \{(\gamma_s, \alpha_s) \mid s = 1, \dots, K\} \tag{5.1.2}$$

is called *Tyurin data* below. This data is related to the moduli of holomorphic vector bundles over Σ. In particular, for generic values of (γ_s, α_s) with $\alpha_s \neq 0$ and $K = ng$ Tyurin data parameterize the semistable rank n degree ng framed holomorphic vector bundles over Σ [75]. In an abuse of the notation, we will omit the subscript s and write just $\{(\gamma, \alpha)\}$ instead $\{(\gamma_s, \alpha_s) \mid s = 1, \dots, K\}$.

Let $M : \Sigma \to \mathfrak{g}$ be a meromorphic function. We require that at a point $\gamma = \gamma_s$ M has the expansion

$$M = \frac{M_{-2}}{(z - z_\gamma)^2} + \frac{M_{-1}}{z - z_\gamma} + M_0 + \cdots, \tag{5.1.3}$$

where z is a fixed local coordinate in the neighborhood of γ, z_γ is the coordinate of γ itself, $M_{-2}, M_{-1}, M_0, M_1, \ldots \in \mathfrak{g}$ and

$$M_{-2} = \lambda\alpha\alpha^t\sigma, \qquad M_{-1} = (\alpha\mu^t + \varepsilon\mu\alpha^t)\sigma, \tag{5.1.4}$$

where $\lambda \in \mathbb{C}$, $\mu \in \mathbb{C}^n$, σ is a $n\times n$-matrix, the upper t denotes the matrix transposition

$$\begin{array}{llll} \lambda \equiv 0, & \varepsilon = 0, & \sigma = id & \text{for } \mathfrak{g} = \mathfrak{gl}(n), \mathfrak{sl}(n), \\ \lambda \equiv 0, & \varepsilon = -1, & \sigma = id & \text{for } \mathfrak{g} = \mathfrak{so}(n), \\ & \varepsilon = 1 & & \text{for } \mathfrak{g} = \mathfrak{sp}(2n), \end{array} \tag{5.1.5}$$

and σ is a matrix of the symplectic form for $\mathfrak{g} = \mathfrak{sp}(2n)$. In the case $\mathfrak{g} = \mathfrak{tsp}(2n)$ we require that α and μ be of the form $\alpha = (\alpha_0, \tilde{\alpha}^t, 0)^t$ ($\alpha_0 \in \mathbb{C}$, $\tilde{\alpha} \in \mathbb{C}^{2n}$), $\mu = (\mu_0, \tilde{\mu}^t, 0)^t$ ($\mu_0 \in \mathbb{C}$, $\tilde{\mu} \in \mathbb{C}^{2n}$), and

$$\sigma = \begin{pmatrix} 0 & 0 & 1 \\ 0 & \boxed{\tilde{\sigma}} & 0 \\ -1 & 0 & 0 \end{pmatrix},$$

where $\tilde{\sigma}$ is the matrix of the symplectic form corresponding to $\mathfrak{sp}(2n)$. For such σ the elements of $\mathfrak{tsp}(2n)$ are of the form

$$X = \begin{pmatrix} 0 & u^t & c \\ 0 & \boxed{\tilde{X}} & v \\ 0 & 0 & 0 \end{pmatrix},$$

where $\tilde{X} \in \mathfrak{sp}(2n)$, $u^t = (a^t, b^t)$, $v^t = (b^t, -a^t)$, and $a, b \in \mathbb{C}^n$. In other words, $\mathfrak{tsp}(2n)$ is the semidirect sum of the $\mathfrak{sp}(2n)$ and the Heisenberg algebra. Here and below we omit the subscripts s, γ indicating the point γ except for z_γ. In the cases $\mathfrak{g} = \mathfrak{sp}(2n)$, $\mathfrak{g} = \mathfrak{tsp}(2n)$ we require that

$$\alpha^t \sigma M_1 \alpha = 0. \tag{5.1.6}$$

Under the above requirements we call M a $\mathfrak{g}$-*valued M-operator*. In [70] we did not require (5.1.6) for M-operators, doing that only for L-operators (see Section 5.2).

Every M-operator defines a tangent vector to the space of Tyurin data by means of derivatives of the coordinates:

$$\partial_M z_\gamma = -\mu^t \sigma \alpha, \quad \partial_M \alpha = -M_0 \alpha + \varkappa \alpha, \tag{5.1.7}$$

where $\varkappa \in \mathbb{C}$ is arbitrary. Since α is given up to the scalar factor, the second equation gives a well-defined tangent vector to the corresponding projective space. Consider M-operators to be functions of $\{(\gamma, \alpha)\}$. Then the relations (5.1.7) define a vector field on T.

Lemma 5.1. *For any two M-operators M_a, M_b and any two vector fields ∂_a, ∂_b on the space of all M-operators satisfying* (5.1.7) *the expression*

$$M_{ab} = \partial_a M_b - \partial_b M_a + [M_a, M_b]$$

is also an M-operator.

The lemma for all Lie algebras from our list is proved in [70], except for $\mathfrak{g} = \mathfrak{sp}(2n), \mathfrak{tsp}(2n)$. We reproduce the proof of that work in order to incorporate the symplectic case. We prove also that M_{ab} satisfies the relation (5.1.6) which was not needed in the setting of [70].

Proof. Let us verify that M_{ab} satisfies (5.1.4).

For an arbitrary $\mathfrak{g}$ from our list we have

$$M_a = \frac{\lambda_a \alpha \alpha^t \sigma}{(z - z_\gamma)^2} + \frac{(\alpha \mu_a^t + \varepsilon \mu_a \alpha^t)\sigma}{z - z_\gamma} + M_{0a} + \dots .$$

and similar expression for M_b where λ_a, λ_b, ε and σ are subjected to (5.1.5) and subsequent remarks. Next we have

$$\begin{aligned}\partial_a M_b &= 2(\partial_a z_\gamma)\frac{\lambda_b \alpha\alpha^t\sigma}{(z-z_\gamma)^3} + \frac{((\partial_a\lambda_b)\alpha\alpha^t + \lambda_b\partial_a(\alpha\alpha^t))\sigma + (\partial_a z_\gamma)M_{-1,b}}{(z-z_\gamma)^2} \\ &+ \frac{((\partial_a\alpha)\mu_b^t + \varepsilon\mu_b(\partial_a\alpha^t) + \alpha(\partial_a\mu_b^t) + \varepsilon(\partial_a\mu_b)\alpha^t)\sigma}{z-z_\gamma} + \cdots\end{aligned} \tag{5.1.8}$$

and similar expression for $\partial_b M_a$.

For the commutator we have

$$\begin{aligned}[M_a, M_b] &= \frac{(1+\varepsilon^2)(\lambda_b\cdot\mu_a^t\sigma\alpha - \lambda_a\cdot\mu_b^t\sigma\alpha)\alpha\alpha^t\sigma}{(z-z_\gamma)^3} \\ &+ \frac{(\lambda_a\partial_b - \lambda_b\partial_a)\alpha\alpha^t\sigma + \lambda_{ab}\alpha\alpha^t\sigma + (\mu_a^t\sigma\alpha)M_{-1,b} - (\mu_b^t\sigma\alpha)M_{-1,a}}{(z-z_\gamma)^2} \\ &+ \frac{((\partial_b\alpha)\mu_a^t + \varepsilon\mu_a(\partial_b\alpha^t))\sigma - ((\partial_a\alpha)\mu_b^t + \varepsilon\mu_b(\partial_a\alpha^t))\sigma}{z-z_\gamma} \\ &+ \frac{(\alpha\mu_{ab}^t + \varepsilon\mu_{ab}\alpha^t)\sigma}{z-z_\gamma} + \cdots\end{aligned} \tag{5.1.9}$$

where $\lambda_{ab} = 2\lambda_b\varkappa_a - 2\lambda_a\varkappa_b + \varepsilon(\mu_a^t\sigma\mu_b - \mu_b^t\sigma\mu_a)$, $\mu_{ab} = \varkappa_a\mu_b - \varkappa_b\mu_a - \lambda_a M_{1b}\alpha + \lambda_b M_{1a}\alpha - M_{0b}\mu_a + M_{0a}\mu_b$. To obtain this relation we used equations (5.1.7) and some additional relations, in particular $\varepsilon\alpha^t\sigma\mu = -\varepsilon^2\mu^t\sigma\alpha$ and $\lambda\alpha^t\sigma\alpha = 0$ which are fulfilled in all cases. In the computation of $[M_a, M_b]_{-2}$ we also used the relation

$$[M_{-1,a}, M_{-1,b}] = (\mu_a^t\sigma\alpha)M_{-1,b} - (\mu_b^t\sigma\alpha)M_{-1,a} + \varepsilon(\mu_a^t\sigma\mu_b - \mu_b^t\sigma\mu_a)\alpha\alpha^t\sigma,$$

which can be verified using (5.1.4). To obtain $[M_a, M_b]_{-1}$ in the form (5.1.9) it is heavily used that $M_{i,a}^t = -\sigma M_{i,a}\sigma^{-1}$ for $\varepsilon \neq 0$ (which follows from (5.1.5)), and the same is true for $M_{i,b}$.

In the case $\mathfrak{g} = \mathfrak{tsp}(2n)$ we need to prove that μ_{ab} has a zero last coordinate. This is obviously the case, because the matrices M_{1a}, M_{1b}, M_{0a}, M_{0b} have zero last rows.

Comparing (5.1.9) and (5.1.8) (and the corresponding relation for $\partial_b M_a$) and using (5.1.7) we obtain

$$M_{ab} = \frac{\tilde\lambda_{ab}\alpha\alpha^t\sigma}{(z-z_\gamma)^2} + \frac{(\alpha\tilde\mu_{ab}^t + \varepsilon\tilde\mu_{ab}\alpha^t)\sigma}{z-z_\gamma} + \cdots,$$

where $\tilde\lambda_{ab} = \partial_a\lambda_b - \partial_b\lambda_a + \lambda_{ab}$, $\tilde\mu_{ab} = \partial_a\mu_b - \partial_b\mu_a + \mu_{ab}$. We observe that M_{ab} has the form (5.1.3), (5.1.4). In particular the -3 order term vanishes because either $\lambda_a = \lambda_b = 0$ or $\varepsilon^2 = 1$ (which follows from (5.1.5)).

Let us prove that M_{ab} satisfies (5.1.6) in the symplectic case. We have

$$(\partial_a M_b)_1 = \partial_a M_{1b} - 2(\partial_a z_\gamma)M_{2b}, \ (\partial_b M_a)_1 = \partial_b M_{1a} - 2(\partial_b z_\gamma)M_{2a}. \quad (5.1.10)$$

Applying ∂_a to both parts of the relation $\alpha^t \sigma M_{1,b}\alpha = 0$, and ∂_b to the corresponding relation for M_{1a}, and using (5.1.7), we obtain

$$\alpha^t \sigma(\partial_a M_{b1})\alpha = -\alpha^t \sigma[M_{0a}, M_{1b}]\alpha, \ \alpha^t \sigma(\partial_b M_{a1})\alpha = -\alpha^t \sigma[M_{0b}, M_{1a}]\alpha. \quad (5.1.11)$$

Then we have

$$[M_a, M_b]_1 = \lambda_a[\alpha\alpha^t\sigma, M_{3b}] + [(\alpha\mu_a^t + \mu_a\alpha^t)\sigma, M_{2b}] + [M_{0a}, M_{1b}] \\ + [M_{1a}, M_{0b}] + [M_{a2}, (\alpha\mu_b^t + \mu_b\alpha^t)\sigma] + \lambda_b[M_{3a}, \alpha\alpha^t\sigma].$$

After taking $\alpha^t\sigma[M_a, M_b]_1\alpha$, the first and the last commutators in the last relation vanish by $\alpha^t\sigma\alpha = 0$ (as well as some terms of the second and the fifth commutators). The two commutators in the middle annihilate with the commutators on the right hand sides of (5.1.11). The remaining terms of the second and the fifth commutators give

$$\alpha^t\sigma(\mu_a\alpha^t\sigma M_{2b} - M_{2b}\alpha\mu_a^t\sigma + M_{2a}\alpha\mu_b^t\sigma - \mu_b\alpha^t\sigma M_{2a})\alpha \\ = 2\alpha^t\sigma\mu_a(\alpha^t\sigma M_{2b}\alpha) - 2\alpha^t\sigma\mu_b(\alpha^t\sigma M_{2a}\alpha)$$

which annihilate with the corresponding terms coming from (5.1.10) due to (5.1.7). □

5.2 L-operators and Lax operator algebras from M-operators

We define here L-operators as M-operators yielding trivial dynamics by (5.1.7). Thus, by definition, every L-operator L is a meromorphic $\mathfrak{g}$-valued function on Σ holomorphic outside $W \cup \{P_+, P_-\}$ such that at a point $\gamma = \gamma_s$

$$L = \frac{L_{-2}}{(z - z_\gamma)^2} + \frac{L_{-1}}{z - z_\gamma} + L_0 + \cdots, \quad (5.2.1)$$

where z is a fixed local coordinate in the neighborhood of γ, z_γ is the coordinate of γ itself, $L_{-2}, L_{-1}, L_0, L_1, \ldots \in \mathfrak{g}$ and

$$L_{-2} = \nu\alpha\alpha^t\sigma, \quad L_{-1} = (\alpha\beta^t + \varepsilon\beta\alpha^t)\sigma \quad (5.2.2)$$

where $\nu \in \mathbb{C}$, $\beta \in \mathbb{C}^n$, σ is a $n\times n$-matrix,

$$\begin{array}{llll} \nu \equiv 0, & \varepsilon = 0, & \sigma = id & \text{for } \mathfrak{g} = \mathfrak{gl}(n), \mathfrak{sl}(n), \\ \nu \equiv 0, & \varepsilon = -1, & \sigma = id & \text{for } \mathfrak{g} = \mathfrak{so}(n), \\ & \varepsilon = 1 & & \text{for } \mathfrak{g} = \mathfrak{sp}(2n), \end{array} \quad (5.2.3)$$

and σ is a matrix of the symplectic form for $\mathfrak{g} = \mathfrak{sp}(2n)$. The case $\mathfrak{g} = \mathfrak{tsp}(2n)$ is not considered in this section; nevertheless, such consideration is possible.

The requirement of triviality of the dynamics (5.1.7) is written as

$$\beta^t \sigma \alpha = 0, \quad L_0 \alpha = \varkappa \alpha. \tag{5.2.4}$$

In addition we assume that

$$\alpha^t \alpha = 0 \text{ for } \mathfrak{g} = \mathfrak{so}(n) \tag{5.2.5}$$

and

$$\alpha^t \sigma L_1 \alpha = 0 \text{ for } \mathfrak{g} = \mathfrak{sp}(2n). \tag{5.2.6}$$

Theorem 5.2. *The space $\mathcal{L} = \overline{\mathfrak{g}}$ of L-operators is a Lie algebra under the point-wise matrix commutator. For $\mathfrak{g} = \mathfrak{gl}(n)$, it is an associative algebra under the point-wise matrix multiplication as well.*

This theorem was proved in Sections 4.1.1–4.1.3 by a straightforward case-by-case computation. In the context of the present chapter, a large part of this computation can be reduced using Lemma 5.1. By that lemma we conclude that for every pair L_a, L_b of L-operators, their bracket $[L_a, L_b]$ is an M-operator and as such is written as (5.1.3) and satisfies (5.1.4), (5.1.5), where $\lambda = \lambda_{ab}$ and $\mu = \mu_{ab}$ are defined by (5.1.9). Interpreting this in terms of L, we obtain the relations (5.2.1)–(5.2.3) for $[L_a, L_b]$ and expressions for the parameters ν, β of that bracket. We use those expressions and the relations (5.2.4)–(5.2.6) for L_a, L_b to prove (5.2.4)–(5.2.6) for their bracket and complete the proof. Note that in the case $\mathfrak{g} = \mathfrak{sp}(2n)$ (5.2.6) is used to prove the first of the relations (5.2.4).

The statement can also be easily proved for $\mathfrak{g} = \mathfrak{tsp}(2n)$. In the later case, making use of the already-proven result in the symplectic case, we obtain that, given $L_1, L_2 \in \overline{\mathfrak{sp}(2n+2)}$, the commutator $[L_1, L_2]$ also belongs to $\overline{\mathfrak{sp}(2n+2)}$. If $L_1, L_2 \in \mathfrak{tsp}(2n)$ at every point then $[L_1, L_2]$ does as well, and according to the proof of the Lemma 5.1, the corresponding β has a zero last coordinate. Hence $[L_1, L_2] \in \overline{\mathfrak{tsp}(2n)}$.

5.3 $\mathfrak{g}$-valued Lax equations

In this section, we consider the consistency of Lax equations of the form

$$L_t = [L, M], \quad L \in \overline{\mathfrak{g}}, \tag{5.3.1}$$

where L, M are an L-operator and an M-operator, respectively. For the purpose of constructing the integrable hierarchy corresponding to equation (5.3.1) (see Section 5.4 below, especially Lemma 5.5) we must assume that M takes values in another Lie algebra $\mathfrak{g}^\diamond$ where the correspondence between $\mathfrak{g}$ and $\mathfrak{g}^\diamond$ is given by

$$\mathfrak{g}^\diamond = \begin{cases} \mathfrak{gl}(n) & \text{if } \mathfrak{g} = \mathfrak{gl}(n), \mathfrak{sl}(n) \\ \mathfrak{so}(2n+1) & \text{if } \mathfrak{g} = \mathfrak{so}(2n), \mathfrak{so}(2n+1) \\ \mathfrak{tsp}(2n) & \text{if } \mathfrak{g} = \mathfrak{sp}(2n). \end{cases} \tag{5.3.2}$$

In all cases $\mathfrak{g}$ is assumed to be embedded into $\mathfrak{g}^\diamond$ in a standard way, and thus the commutator $[L, M]$ makes sense.

Following [33] let us assign every effective divisor $D = \sum_i m_i P_i$ on Σ with the space $\mathcal{L}^D = \bigcup_{(\alpha,\gamma)} L^D_{\alpha,\gamma}$ (join over all Tyurin parameters (α, γ) where

$$L^D_{\alpha,\gamma} = \{L \in \overline{\mathfrak{g}}_{\alpha,\gamma} \mid (L) + D \geq 0\}$$

and $\overline{\mathfrak{g}}_{\alpha,\gamma}$ is the Lax operator algebra corresponding to (α, γ).

Under certain conditions given by Lemma 5.4 below, the Lax equation (5.3.1) gives a well-defined dynamical system on $\mathcal{L}^D$.

Let the upper dot mean the time derivative and the term *γ-points* be reserved for the points γ_s.

Lemma 5.3. *At γ-points, equations* (5.3.1) *imply the following equations:*

$$\dot{z}_\gamma = -\mu^t \sigma \alpha, \quad \dot{\alpha} = -M_0 \alpha + \varkappa \alpha. \tag{5.3.3}$$

For $\mathfrak{g} = \mathfrak{so}(n)$ the statement is true in a generic position (specified in the proof).

Proof. By a straightforward computation we have

$$\begin{aligned} \dot{L} = 2\dot{z}_\gamma \frac{\nu\alpha\alpha^t\sigma}{(z - z_\gamma)^3} &+ \frac{\dot{\nu}\alpha\alpha^t\sigma + \nu\dot{\alpha}\alpha^t\sigma + \nu\alpha\dot{\alpha}^t\sigma + \dot{z}_\gamma(\alpha\beta^t + \varepsilon\beta\alpha^t)\sigma}{(z - z_\gamma)^2} \\ &+ \frac{\dot{\alpha}\beta^t\sigma + \alpha\dot{\beta}^t\sigma + \varepsilon\dot{\beta}\alpha^t\sigma + \varepsilon\beta\dot{\alpha}^t\sigma}{(z - z_\gamma)} + (\dot{L}_0 - \dot{z}_\gamma L_1) + \cdots. \end{aligned} \tag{5.3.4}$$

For $\mathfrak{g} = \mathfrak{gl}(n)$ we must take $\nu = 0$, $\varepsilon = 0$. Looking for a solution to (5.3.1) in Laurent expansions at γ we obtain $\dot{z}_\gamma = -\mu^t\alpha$ in the order -2 in $z - z_\gamma$, and

$$\dot{\alpha}\beta^t + \alpha\dot{\beta}^t = \alpha\beta^t M_0 - M_0\alpha\beta^t + \varkappa\alpha\mu^t - \alpha\mu^t L_0$$

in the order -1. Let us multiply both parts of the last relation by a $\tilde{\beta}$ which is not orthogonal to β. We obtain

$$\dot{\alpha}\beta^t\tilde{\beta} + \alpha\dot{\beta}^t\tilde{\beta} = \alpha\beta^t M_0\tilde{\beta} - M_0\alpha\beta^t\tilde{\beta} + \varkappa\alpha\mu^t\tilde{\beta} - \alpha\mu^t L_0\tilde{\beta}.$$

The latter can be resolved with respect to $\dot{\alpha}$, giving

$$\dot{\alpha} = -M_0\alpha + \tilde{\varkappa}\alpha, \ \tilde{\varkappa} \in \mathbb{C}. \tag{5.3.5}$$

In the case $\mathfrak{g} = \mathfrak{so}(n)$, we take $\tilde{\beta}$ to be orthogonal to α^t and $\dot{\alpha}^t - \alpha^t M_0$, and the proof is similar. Since $\tilde{\beta}$ should not be orthogonal to β, we must require that β d not equal any linear combination of α and $\dot{\alpha} - M_0\alpha$, which is a condition of a generic position.

In the case $\mathfrak{g} = \mathfrak{sp}(2n)$ the relation $\dot{z}_\gamma = -\mu^t\sigma\alpha$ comes from the comparison of the terms of the -3 order, and the relation (5.3.5) comes of the order -2 terms. In the last case we obtain

$$\begin{aligned}&\alpha\alpha^t\sigma + \dot{\alpha}\alpha^t\sigma + \alpha\dot{\alpha}^t\sigma + \dot{z}_\gamma(\alpha\beta^t + \beta\alpha^t) = \\ &\quad \alpha\alpha^t\sigma M_0 - M_0\alpha\alpha^t\sigma + \alpha\beta^t\sigma\mu\alpha^t\sigma - \alpha\mu^t\sigma\beta\alpha^t\sigma + \alpha(2\varkappa\alpha^t\sigma) + \dot{z}_\gamma(\alpha\beta^t + \beta\alpha^t).\end{aligned}$$

It is easily resolvable with respect to $\dot{\alpha}$ by the same trick (one takes a $\tilde{\beta}$ such that $\alpha^t\sigma\beta \neq 0$). □

Equations (5.3.3) are simply the equations of motion for Tyurin parameters. They are heavily used in [33]. Originally, the concept of moving Tyurin parameters was invented in [25], where it was used for an effective solution of Kadomtsev–Petviashvili equations in certain cases.

Let $T_L\mathcal{L}^D$ denote the tangent space to $\mathcal{L}^D$ at a point L.

Lemma 5.4. $[L, M] \in T_L\mathcal{L}^D \Leftrightarrow ([L, M]) + D \geq 0$ *outside* γ*'s, the provided equations* (5.3.3) *are fulfilled at every* γ.

Proof. In our proof we follow the lines of [33], where the lemma was formulated and proved for $\mathfrak{g} = \mathfrak{gl}(n)$.

Let z be a local coordinate in an open set containing a point γ, and z_γ be the corresponding coordinate of γ.

Identify $T\mathcal{L}^D$ with the space $\mathcal{T}^D$ of all meromorphic $\mathfrak{g}$-valued functions T such that at every γ

$$\begin{aligned}T = 2\dot{z}_\gamma\frac{\nu\alpha\alpha^t\sigma}{(z-z_\gamma)^3} &+ \frac{\dot{\nu}\alpha\alpha^t\sigma + \nu(\dot{\alpha}\alpha^t + \alpha\dot{\alpha}^t)\sigma + \dot{z}_\gamma(\alpha\beta^t + \varepsilon\beta\alpha^t)\sigma}{(z-z_\gamma)^2} \\ &+ \frac{(\dot{\alpha}\beta^t + \varepsilon\beta\dot{\alpha}^t + \alpha\dot{\beta}^t + \varepsilon\dot{\beta}\alpha^t)\sigma}{z-z_\gamma} + T_0 + \cdots\end{aligned} \tag{5.3.6}$$

$$\dot{\alpha}^t\sigma\beta + \alpha^t\sigma\dot{\beta} = 0 \tag{5.3.7}$$

$$T_0\alpha = \varkappa\dot{\alpha} + \dot{\varkappa}\alpha - L_0\dot{\alpha} - \dot{z}_\gamma L_1\alpha \tag{5.3.8}$$

where $\dot{z}_\gamma, \dot{\varkappa}$ are constants, $\dot{\alpha}, \dot{\beta}$ are constant vectors fulfilling the relations (5.3.3), and the divisor of T outside the points γ is greater or equal to $-D$.

The relation (5.3.6) is modelled on (5.3.4) which is obtained by the time derivation of (5.2.1). In particular

$$T_0 = \dot{L}_0 - \dot{z}_\gamma L_1.$$

Together with the time derivation of (5.2.4) this gives (5.3.8). Thus $T\mathcal{L}^D$ embeds to $\mathcal{T}^D$. Let us check the coincidence of the dimensions of those spaces. It can be done in a quite uniform way for all Lie algebras, and so we show it in the most difficult case $\mathfrak{g} = \mathfrak{sp}(2n)$. We have $(T) + \tilde{D} \geq 0$ where $\tilde{D} = D + 3\sum \gamma$, and $\deg \tilde{D} = \deg D + 3K$. By the Riemann–Roch theorem, $\dim\{T|(T) + \tilde{D} \geq 0\} = (\dim \mathfrak{g})(\deg D + 3K - g + 1)$. The elements of $\mathcal{T}^D$ are distinguished in the space $\{T|(T) + \tilde{D} \geq 0\}$ by the following relations. First, at every point γ we have

$$\begin{aligned} T_{-3} &= 2\dot{z}_\gamma \nu\alpha\alpha^t\sigma, \\ T_{-2} &= \dot{\nu}\alpha\alpha^t\sigma + \nu(\dot{\alpha}\alpha^t + \alpha\dot{\alpha}^t)\sigma + \dot{z}_\gamma(\alpha\beta^t + \beta\alpha^t)\sigma, \\ T_{-1} &= (\dot{\alpha}\beta^t + \alpha\dot{\beta}^t + \dot{\beta}\alpha^t + \beta\dot{\alpha}^t)\sigma. \end{aligned} \tag{5.3.9}$$

Since the elements on the left-hand side belong to $\mathfrak{g}$, (5.3.9) gives $3\dim\mathfrak{g}$ relations. Taking account of (5.3.7) and (5.3.8) gives $2n+1$ relations, and (5.3.3) gives another $2n+1$ relation. Thus we have $3\dim\mathfrak{g}+4n+2$ relations. These relations contain $4n+2$ free parameters $\dot{z}_\gamma, \dot{\nu}, \dot{\alpha}, \dot{\beta}$. Thus we actually obtained $3\dim\mathfrak{g}$ relations at every point γ, and the number of those points is K; hence we have $3(\dim\mathfrak{g})K$ relations. We see that $\dim \mathcal{T}^D = (\dim\mathfrak{g})(\deg D - g + 1)$.

But $\mathcal{L}^D$ has the same dimension. We can count this in a quite similar way or make use of Theorem 4.4. Assume that we are in the two-point situation, i.e., $D = -m_+P_+ + (m_- + g)P_-$ where $m_- > m_+$ for simplicity. Then $\mathcal{L}^D = \mathfrak{g}_{m_+} \oplus \dots \oplus \mathfrak{g}_{m_-}$. By Theorem 4.4, $\dim\mathcal{L}^D = (\dim\mathfrak{g})(m_- - m_+ + 1)$, which is exactly equal to $(\dim\mathfrak{g})(\deg D - g + 1)$. We conclude that $\dim \mathcal{T}^D = \dim T_L\mathcal{L}^D$, and hence these linear spaces coincide.

Next we prove that if L, M are as above, then $[L, M]$ possesses the properties (5.3.6)–(5.3.8), i.e., belongs to $\mathcal{T}^D$. The proof is once again straightforward. For example, show (5.3.8). Denote the degree zero term $[L, M]_0$ of the commutator by T_0. Then in the case $\mathfrak{g} = \mathfrak{gl}(n)$ we find by a computation

$$T_0\alpha = \alpha(\beta^t M_1\alpha - \mu^t L_1\alpha) + (L_0 - \varkappa)M_0\alpha + L_1\alpha(\mu^t\alpha). \tag{5.3.10}$$

If we replace $\mu^t\alpha$ with $-\dot{z}_\gamma$, $M_0\alpha$ with $-\dot{\alpha}$ and denote $\beta^t M_1 - \mu^t L_1\alpha$ by $\dot{\varkappa}$, we obtain (5.3.8). For other types of $\mathfrak{g}$ the expression for $T_0\alpha$ is more complicated, and we use the relations (5.2.4)–(5.2.6) to identify it with (5.3.8).

In the case $\mathfrak{g} = \mathfrak{sp}(2n)$ we get

$$\begin{aligned} T_0\alpha = {} & \alpha(\nu\alpha^t\sigma M_2\alpha + \beta^t\sigma M_1\alpha - \mu^t\sigma L_1\alpha - \lambda\alpha^t\sigma L_2\alpha) + (L_0 - \varkappa)M_0\alpha \\ & + L_1\alpha(\mu^t\sigma\alpha) + \beta\alpha^t\sigma M_1\alpha - \mu\alpha^t\sigma L_1\alpha \end{aligned}$$

instead of (5.3.10). Making use of the relations $\alpha^t\sigma M_1\alpha = 0$, $\alpha^t\sigma L_1\alpha = 0$ we obtain the same result. □

Lemma 5.4 directly implies that whenever $([L, M]) + D \geq 0$ outside γ's and the equations of moving poles are fulfilled, the Lax equation (5.3.1) is consistent.

5.4 Hierarchies of commuting flows

In the case of $\mathfrak{g} = \mathfrak{gl}(n)$ I. Krichever [33] has shown that the Lax operator considered as a function on $\mathcal{L}^D$ yields the hierarchy of commuting flows on $\mathcal{L}^D$. Here we present a generalization of that result on the classic Lie algebras. Recall from the last section that M takes values in the Lie algebra $\mathfrak{g}^\diamond$ given by (5.3.2).

For a divisor $D = \sum m_i P_i$ define a divisor $\widetilde{D} = D + \delta \sum_{s=1}^{K} \gamma_s$ where

$$K = \begin{cases} ng, & \mathfrak{g} = \mathfrak{gl}(n),\ \mathfrak{sl}(n),\ \mathfrak{so}(2n),\ \mathfrak{so}(2n+1), \\ (n+1)g, & \mathfrak{g} = \mathfrak{sp}(2n) \end{cases}$$

and

$$\delta = \begin{cases} 1, & \mathfrak{g} = \mathfrak{gl}(n),\ \mathfrak{sl}(n),\ \mathfrak{so}(2n),\ \mathfrak{so}(2n+1), \\ 2, & \mathfrak{g} = \mathfrak{sp}(2n), \end{cases}$$

(K and δ actually depend on $\mathfrak{g}^\diamond$).

Let us define $\mathcal{N}^D$ as a space of M-operators such that $(M) + \widetilde{D} \geq 0$.

Lemma 5.5. $\dim \mathcal{N}^D = (\dim \mathfrak{g}^\diamond)(\deg D + 1)$.

Proof. We compute $\dim \mathcal{N}^D$ by the Riemann–Roch theorem taking into account the additional relations at the points γ. These relations determe M_{-2}, M_{-1}, and the constraint (5.1.6) on M_1 (the latter only in the case $\mathfrak{g} = \mathfrak{sp}(2n)$). At every point γ the number of the relations is equal to $\delta \dim \mathfrak{g}^\diamond$ if $\mathfrak{g} \neq \mathfrak{sp}(2n)$, and is equal to $\delta \dim \mathfrak{g}^\diamond + 1$ in the last case. We write that number in the form $\delta \dim \mathfrak{g}^\diamond + r_{\mathfrak{sp}}$ where $r_{\mathfrak{sp}} = 1$ for $\mathfrak{g} = \mathfrak{sp}(2n)$ and $r_{\mathfrak{sp}} = 0$ otherwise. We also have free parameters μ, λ (λ appears only in the case $\mathfrak{g} = \mathfrak{sp}(2n)$). Let r be the number of those parameters for a fixed γ. We can think that there are $\delta \dim \mathfrak{g} - r + r_{\mathfrak{sp}}$ relations at every γ.

Let us write K in the form $K = lg$, where l is equal to n or to $n+1$ depending on $\mathfrak{g}^\diamond$. We have

$$\begin{aligned} \dim N^D &= (\dim \mathfrak{g}^\diamond)(\deg D + \delta l g - g + 1) - (\delta \dim \mathfrak{g}^\diamond - r + r_{\mathfrak{sp}}) l g \\ &= (\dim \mathfrak{g}^\diamond)(\deg D + 1) - \big(\dim \mathfrak{g}^\diamond - (r - r_{\mathfrak{sp}}) l\big) g. \end{aligned} \tag{5.4.1}$$

Next verify that

$$\dim \mathfrak{g}^\diamond = (r - r_{\mathfrak{sp}}) l. \tag{5.4.2}$$

Recall from Section 5.3 that there are three cases: $\mathfrak{g}^\diamond = \mathfrak{gl}(n)$, $\mathfrak{g}^\diamond = \mathfrak{so}(2n+1)$, and $\mathfrak{g}^\diamond = \mathfrak{sp}(2n)$. Indeed, for $\mathfrak{g}^\diamond = \mathfrak{gl}(n)$ we have $r = l = n$, $r_{\mathfrak{sp}} = 0$, and hence $(r - r_{\mathfrak{sp}}) l = n^2$. If $\mathfrak{g}^\diamond = \mathfrak{so}(2n+1)$, then $r = 2n + 1$, $l = n$, $r_{\mathfrak{sp}} = 0$, and

$(r - r_{\mathfrak{sp}})l = (2n+1)n$. Finally, if $\mathfrak{g}^\diamond = \mathfrak{tsp}(2n)$, then $\dim \mathfrak{g}^\diamond = n(2n+1) + (2n+1) = (2n+1)(n+1)$ (the sum of the dimensions of $\mathfrak{sp}(2n)$ and the Heisenberg algebra; see page 104). We have $r = 2n+2$ in this case ($2n+1$ free parameters come from μ and 1 corresponds to λ). Hence $r - r_{\mathfrak{sp}} = 2n+1$ and $(r - r_{\mathfrak{sp}})l = (2n+1)(n+1)$.

In all cases (5.4.2) is true. □

Following [33] let us fix a point $P_0 \in \Sigma$ and local coordinates w_0, w_i in the neighborhood of the points P_0, P_i. Our next goal is to define gauge invariant functions M_a that satisfy the assumptions of Lemma 5.4. Let us define a as a triple:

$$a = (P_i, k, m), \quad k > 0, \quad m > -m_i, \tag{5.4.3}$$

where k, m are integers, $k \equiv 1 (\mathrm{mod}\, 2)$ for $\mathfrak{g}^\diamond = \mathfrak{so}(2n+1)$, and $\mathfrak{g}^\diamond = \mathfrak{sp}(2n)$.

By Lemma 5.5 for generic $L \in \overline{\mathfrak{g}}$ there is a unique $\mathfrak{g}^\diamond$-valued meromorphic function M_a such that

(i) M_a is an M-operator;
(ii) outside the points γ it has a pole only at the point P_i, and

$$M_a(w) = w_i^{-m} L^n(w) + O(1),$$

i.e., singular parts of M_a and $w_i^{-m} L^n$ coincide;
(iii) M_a is normalized by the condition $M_a(P_0) = 0$.

Theorem 5.6. *The equations*

$$\partial_a L = [L, M_a], \quad \partial_a = \partial/\partial t_a \tag{5.4.4}$$

define a hierarchy of commuting flows on an open subset of $\mathcal{L}^D$.

For $\mathfrak{g} = \mathfrak{gl}(n)$ the theorem is formulated and proved in [33]. For $\mathfrak{g} = \mathfrak{so}(2n)$, $\mathfrak{so}(2n+1)$ it is proved in [70] under slightly different assumptions, and for $\mathfrak{g} = \mathfrak{sp}(2n)$ in [71].

Proof. It follows from (ii) that $([L, M_a]) + D \geq 0$, hence by Lemma 5.4 $[L, M_a] \in T_L \mathcal{L}^D$ and the equation $\partial_a L = [L, M_a]$ defines a flow on $\mathcal{L}^D$.

To prove commutativity of such flows, it is sufficient to verify that $M_{ab} = \partial_a M_b - \partial_b M_a + [M_a, M_b] = 0$ identically. First, by Lemma 5.3 the vector fields ∂_a, ∂_b satisfy the relations (5.1.7). Hence by Lemma 5.1 M_{ab} is an M-operator. Below we prove that this M-operator is regular at the points of the divisor D. By Lemma 5.5 the space of such operators has the same dimension as $\mathfrak{g}^\diamond$. Due to (iii) we obtain $M_{ab} = 0$.

Let us prove that M_{ab} is regular at the points of the divisor D. We repeat here the corresponding part of the proof of [33, Theorem 2.1]. First, assume that indices

a,b correspond to the same point P_i, i.e., $a = (P_i, n, m)$, $b = (P_i, n', m')$. Denote $M_a - w^{-m}L^n$ by M_a^- and $M_b - w^{-m'}L^{n'}$ by M_b^-, then by (ii) M_a^- and M_b^- are regular in the neighborhood of P_i. We have

$$\begin{aligned}\partial_a M_b = w^{-m'}\partial_a L^{n'} + \partial_a M_b^- = w^{-m'}[L^{n'}, M_a] + \partial_a M_b^- \\ - w^{-m'}[L^{n'}, M_a^-] + \partial_a M_b^-\end{aligned}$$

and

$$\begin{aligned}[M_a, M_b] &= [M_a^- + w^{-m}L^n, M_b^- + w^{-m'}L^{n'}] \\ &= w^{-m}[L^n, M_b^-] - w^{-m'}[L^{n'}, M_a^-] + [M_a^-, M_b^-].\end{aligned}$$

Hence $M_{ab} = \partial_a M_b^- - \partial_b M_a^- + [M_a^-, M_b^-]$ at the point P_i, which is a regular expression at that point. By definition M_{ab} is regular also at the other points of D.

The proof is similar in the case when a and b correspond to the different points of D. □

5.5 Symplectic structure

Following the lines of [33] we introduce here a symplectic structure on a certain subspace $\mathcal{P}^D \subset \mathcal{L}^D/G$ where $G = \exp \mathfrak{g}$. We call it the *Krichever–Phong symplectic structure*.

Let Ψ be the matrix formed by the canonically normalized left eigenvectors of L. In the case $\mathfrak{g} = \mathfrak{gl}(n)$ following [33] we consider a vector ψ to be canonically normalized if $\sum \psi_i = 1$. In the other cases we require that $\Psi \in G$ point-wise, where $G = SO(n), Sp(2n)$ depending on $\mathfrak{g}$, i.e., $\Psi^t = -\varepsilon\sigma\Psi^{-1}\sigma^{-1}$, and ε satisfies to $\sigma^t = -\varepsilon\sigma$, i.e., $\varepsilon = -1$ if σ is symmetric (the case $\mathfrak{g} = \mathfrak{so}(n)$), $\varepsilon = 1$ if σ is skew-symmetric (the case $\mathfrak{g} = \mathfrak{sp}(2n)$).

The Ψ is defined modulo permutations of its rows. We consider L and Ψ as matrix-valued functions on $\mathcal{L}^D$. Let δL and $\delta\Psi$ denote their external differentials, which are 1-forms on $\mathcal{L}^D$. In the same way we consider the diagonal matrix K defined by

$$\Psi L = K\Psi,$$

i.e., formed by the eigenvalues of L, and the matrix-valued 1-form δK. Let Ω be a 2-form on $\mathcal{L}^D$ with values in the space of meromorphic functions on Σ defined by the relation

$$\Omega = \operatorname{tr}(\delta\Psi \wedge \delta L \cdot \Psi^{-1} - \delta K \wedge \delta\Psi \cdot \Psi^{-1}).$$

Ω does not depend on the order of the eigenvalues, hence it is well defined on $\mathcal{L}$.

Fix a holomorphic differential ϖ on Σ and define a scalar-valued 2-form ω on $\mathcal{L}^D$ by the relation

$$\omega = -\frac{1}{2}\Big(\sum_{s=1}^{K} \operatorname{res}_{\gamma_s} \Omega\varpi + \sum_{P_i\in D} \Omega\varpi\Big).$$

There is another representation for Ω:

$$\Omega = 2\delta \operatorname{tr}(\delta\Psi \cdot \Psi^{-1} K),$$

which implies that ω is apparently closed. First we want to prove that it is nondegenerate when restricted to the space of Tyurin parameters, i.e., ω yields a symplectic form on this space. We will point out a canonical form of that restriction.

Lemma 5.7. *The contribution of Tyurin parameters to ω is of the form*

$$\omega_0 = \sum_{s=1}^{K} (a\,\delta z_s \wedge \delta\varkappa_s + \delta\beta_s^t \wedge \delta\alpha_s)$$

where $a = 1$ for $\mathfrak{g} = \mathfrak{gl}(n)$, $a = 2$ for $\mathfrak{g} = \mathfrak{so}(n)$ and $\mathfrak{g} = \mathfrak{sp}(2n)$.

Proof. For $\mathfrak{g} = \mathfrak{gl}(n)$ the corresponding statement is contained in Lemmas 2.1 and 4.3 [33]. It is instructive to reproduce the proof here. Let g_s be a constant nondegenerate matrix such that $g_s\alpha = e_1$ where $e_1^t = (1, 0, \ldots, 0)$. Then for $L_s' = g_s L_s g_s^{-1}$ we have $(L_s')_{-1} = e_1 f^t$ where $f^t = \beta^t g_s^{-1}$, and $f_1 = 0$ since $f^t e_1 = 0$. Apparently, only the entries of the first row of the matrix $L_{s,-1}'$ are nonzero, but $(L_{s,-1}')_{11} = 0$.

The e_1 is an eigenvector for L_{s0} with the eigenvalue $\varkappa$. For that reason $(L_{s0})^{11} = \varkappa$, $(L_{s0})^{i1} = 0$ $(i > 1)$.

We can conclude from the above remarks that the conjugation by the matrix $f_s = \operatorname{diag}(z - z_s, 1, \ldots, 1)$ takes $(z - z_s)^{-1} L_{-1}$ and L_0 to holomorphic matrix-valued functions. Hence the same is true for L_s', and the lemma is proven with $\Phi_s = f_s g_s$. Let us note for the future that f_s is diagonal.

Let us watch now for the transformations of Ω corresponding to the transformations of L just performed.

Under the gauge transformation $L' = gLg^{-1}$, $\Psi' = \Psi g^{-1}$, the form Ω transforms to Ω', where $\Omega' = \Omega + 2\operatorname{tr}(\delta L \wedge g^{-1}\delta g - Lg^{-1}\delta g \wedge g^{-1}\delta g)$.

For $g = g_s$ we obtain

$$\operatorname{res}_{\gamma_s} \Omega'\varpi = \operatorname{res}_{\gamma_s} \Omega\varpi + 2\operatorname{res}_{\gamma_s} \operatorname{tr}(\delta L \wedge g_s^{-1}\delta g_s - Lg_s^{-1}\delta g_s \wedge g_s^{-1}\delta g_s)\varpi.$$

Since g_s is constant we have

$$\operatorname{res}_{\gamma_s} \Omega'\varpi = \\ \operatorname{res}_{\gamma_s} \Omega\varpi - 2\operatorname{res}_{\gamma_s} \operatorname{tr}\big((\delta\alpha_s \cdot \beta_s^t + \alpha_s \cdot \delta\beta_s^t) \wedge \delta g_s g_s^{-1} + \alpha_s\beta_s^t \delta g_s g_s^{-1} \wedge \delta g_s g_s^{-1}\big).$$

By differentiating the relation $g_s\alpha = e_1$ we obtain $\delta\alpha = -g_s^{-1}\delta g_s\alpha$. Substituting this to the previous relation we obtain

$$\operatorname{res}_{\gamma_s} \Omega'\varpi = \operatorname{res}_{\gamma_s} \Omega\varpi + 2\operatorname{tr}(\delta\alpha_s \wedge \delta\beta_s^t).$$

The matrix L'_s becomes holomorphic under the transformation $\widehat{L}_s = f_s L'_s f_s^{-1}$, hence

$$0 = \operatorname{res}_{\gamma_s} \Omega'\varpi + 2\operatorname{res}_{\gamma_s} \operatorname{tr}(\delta L' \wedge f_s^{-1}\delta f_s - L'\delta f_s^{-1} f_s \wedge \delta f_s^{-1} f_s).$$

As f_s is diagonal the last term vanishes. Making use of the above obtained special form of L' and f_s we conclude that

$$\operatorname{res}_{\gamma_s} \Omega'\varpi = -2\delta z_s \wedge \delta\varkappa_s.$$

Thus the contribution of the point γ_s to Ω is equal to

$$\operatorname{res}_{\gamma_s} \Omega = -2\operatorname{tr}(\delta\alpha_s \wedge \delta\beta_s^t) - 2\delta z_s \wedge \delta\varkappa_s,$$

and the corresponding contribution to ω is equal to

$$\omega_s = -\frac{1}{2}\operatorname{res}_{\gamma_s} \Omega = \delta\beta_s^t \wedge \delta\alpha_s + \delta z_s \wedge \delta\varkappa_s.$$

The proof for $\mathfrak{so}(n)$ and $\mathfrak{sp}(2n)$ is more complicated but similar. In the symplectic case we choose

$$\sigma = \begin{pmatrix} \sigma' & 0 \\ 0 & \sigma'' \end{pmatrix} \quad \text{where} \quad \sigma' = \begin{pmatrix} 0 & 1 \\ -1 & 0 \end{pmatrix}$$

as a matrix giving the symplectic form.

We can send any α to any other vector by a symplectic transformation g. Let us send α to e_1 where $e_1^t = (1, 0, \dots, 0)^t$. Then

$$\alpha\alpha^t\sigma = \begin{pmatrix} 0 & 1 & 0 & \dots & 0 \\ 0 & 0 & 0 & \dots & 0 \\ 0 & 0 & 0 & \dots & 0 \\ \vdots & \vdots & \vdots & \ddots & \vdots \\ 0 & 0 & 0 & \dots & 0 \end{pmatrix}, \quad (\alpha\beta^t + \beta\alpha^t)\sigma = \begin{pmatrix} -\beta_2 & 2\beta_1 & * & \dots & * \\ 0 & \beta_2 & 0 & \dots & 0 \\ 0 & \beta_3 & 0 & \dots & 0 \\ \vdots & \vdots & \vdots & \ddots & \vdots \\ 0 & \beta_{2n} & 0 & \dots & 0 \end{pmatrix},$$

where the stars denote constants linearly depending on β. Observe that by $\beta^t\sigma\alpha = 0$ we have $\beta_2 = 0$.

It is easy to check that conjugation by $f = \begin{pmatrix} Z & 0 \\ 0 & E \end{pmatrix}$, where $Z = \begin{pmatrix} z & 0 \\ 0 & z^{-1} \end{pmatrix}$, multiplies the only nonzero entry of $\alpha\alpha^t\sigma$ by z^2, and the matrix $(\alpha\beta^t + \beta\alpha^t)\sigma$ by z (except for the entry $2\beta_1$ multiplied by z^2). Hence $f \cdot (\frac{\alpha\alpha^t\sigma}{z^2} + \frac{(\alpha\beta^t+\beta\alpha^t)\sigma}{z}) \cdot f^{-1}$ is holomorphic.

The matrix L_0 is similar to $(\alpha\beta^t + \beta\alpha^t)\sigma$ in structure. Its first column is equal to $(\varkappa, 0, \dots, 0)^t$ by $L_0 e_1 = \varkappa e_1$. Its second row is equal to $(0, -\varkappa, \dots, 0)^t$ by $(e_1^t\sigma)L_0 = -\varkappa(e_1^t\sigma)$ (which follows from $L_0^t\sigma + \sigma L_0 = 0$). Nonzero entries of L_0 are multiplied by positive degrees of z under the conjugation by f, except for the right lower $(2n-2)\times(2n-2)$ corner block which is invariant. Hence fL_0f^{-1} is holomorphic.

A singularity could only come from $f\cdot(zL_1)f^{-1}$, because $(L_1)_{21}$ is multiplied by z^{-2}. But $\alpha^t\sigma L_1\alpha = 0$ implies $(L_1)_{21} = 0$. All other entries of $f\cdot L_1 f^{-1}$ are of the order z^{-1} and more, hence when multiplied by z they become holomorphic.

The matrix $\delta f\cdot f^{-1}$ has null entries except for the left upper 2×2 corner block which is equal to $\begin{pmatrix}\delta z\cdot z^{-1} & 0\\ 0 & -\delta z\cdot z^{-1}\end{pmatrix}$. As it follows from what was said of the structure of L_0, the corresponding block of δL_0 is equal to $\begin{pmatrix}\delta\varkappa & *\\ 0 & -\delta\varkappa\end{pmatrix}$. Thus, at a point γ_s, we have the contribution $\delta\beta_s^t\wedge\delta\alpha_s$ from the gauge transformation corresponding to g, and $2\delta z_s\wedge\delta\gamma_s$ from the gauge transformation corresponding to f. The total contribution of a point γ_s to ω is equal to $2\delta z_s\wedge\delta\varkappa_s + \delta\beta_s^t\wedge\delta\alpha_s$.

In the orthogonal case we must additionally satisfy the relation $e_1^t\sigma e_1 = 0$. Consider $\mathfrak{g} = \mathfrak{so}(2n)$ for example. Let us choose σ as in the symplectic case but take $\sigma' = \begin{pmatrix}0 & 1\\ 1 & 0\end{pmatrix}$ and assume σ'' to be positive definite. Then e_1 (where $e_1^t = (1, 0, \dots, 0)^t$) satisfies the required relation and we proceed as in the symplectic case, with the same matrix function f. The structure of $L_{-1} = (\alpha\beta^t - \beta\alpha^t)\sigma$ is similar to that in the symplectic case, almost the only difference being that the left upper corner 2×2-block is of the form $\begin{pmatrix}\beta_2 & *\\ 0 & -\beta_2\end{pmatrix}$. This difference is technical and does not affect the result. □

Let us now consider the contribution of the marked points $P_1, \dots, P_N$. Let $\omega_m = -\frac{1}{2}\operatorname{res}_{P_m}\Omega\varpi$. Define $\mathcal{P}_0^D$ as a subspace in $\mathcal{L}^D$ where the 1-form $\delta\varkappa\,\varpi$ is holomorphic. It is the same as the set of common zeroes of the following functions on $\mathcal{L}^D$:

$$T_{ijl} = \operatorname{res}_{P_i^l}\left(\left(z - z(P_i)\right)^j \varkappa\,\varpi\right), \quad j = 0, \dots, (m_i - d_i),$$

where l labels the sheets of the spectral curve (as a branch cover of Σ), $d_i = \operatorname{ord}_{P_i}\varpi$, and z is a local coordinate in the neighborhood of P_i. Note that the foliation given by the common level sets of those functions is invariant with respect to the flows of our commuting hierarchy, since they preserve the spectrum $\varkappa$ (see also the next section).

Lemma 5.8. *On the space* $\mathcal{P}_0^D$

$$\omega_m = \operatorname{res}_{P_m}\operatorname{tr}(L\,\Psi^{-1}\delta\Psi\wedge\Psi^{-1}\delta\Psi)\varpi. \tag{5.5.1}$$

Proof. By differentiating the relation $L = \Psi^{-1}K\Psi$ we exclude δL in the definition of Ω and obtain

$$\omega_m = \operatorname{res}_{P_m}\operatorname{tr}(K\,\delta\Psi\cdot\Psi^{-1}\wedge\delta\Psi\cdot\Psi^{-1} - \delta K\wedge\delta\Psi\cdot\Psi^{-1})\varpi.$$

The second summand is holomorphic at P_m on the space $\mathcal{P}_0^D$, because such is $\delta K\varpi$, and the form $\delta\Psi \cdot \Psi^{-1}$ is always holomorphic, since $\delta\Psi$ and Ψ have the same order at P_m (the point P_m is immovable and there is no variation in the local coordinate on Σ while taking $\delta\Psi$). For the first summand we have $\operatorname{tr}(K\,\delta\Psi \cdot \Psi^{-1} \wedge \delta\Psi \cdot \Psi^{-1}) = \operatorname{tr}(L\,\Psi^{-1}\delta\Psi \wedge \Psi^{-1}\delta\Psi)$. □

Let ξ be a tangent vector to $\mathcal{P}_0^D$ at a point L. It is a $\mathfrak{g}$-valued meromorphic function on Σ, and $\delta\Psi(\xi) \cdot \Psi^{-1} = \xi$. Hence for any pair ξ, η of tangent vectors we have $\omega_m(\xi, \eta) = \operatorname{res}_{P_m} \operatorname{tr}(L[\xi, \eta])$. This is a 2-form of the Kirillov type. In general it is degenerate and has symplectic leaves. In particular, in the case when $L\varpi$ has a simple pole at P_m with the residue L_m, the ω_m descends to the canonical Kirillov form on the orbit O_m of the element $L_m \in \mathfrak{g}$. It is obvious that $(L\varpi) + D - (\varpi) \geq 0$ for $L \in \mathcal{L}^D$ where (ϖ) is a zero divisor of the form ϖ. Contrary to the previous case, if some components of the divisor $D - (\varpi)$ are negative, the corresponding ω_m vanish, and ω obviously degenerates. We omit the details and refer to the complete analogy with [33] in this respect.

Since ω is G-invariant, it is actually defined on $\mathcal{P}^D = \mathcal{P}_0^D/G$. Together with what has been already proven in this section, we have the following theorem, which is quite similar to [33, Theorem 4.1].

Theorem 5.9. *If $D - (\varpi) \geq 0$, then the form*

$$\omega = \sum_s (a\,\delta\varkappa_s \wedge \delta z_s + \delta\alpha_s^t \wedge \delta\beta_s) + \sum_m \omega_m. \tag{5.5.2}$$

is nondegenerate on $\mathcal{P}^D$. Thus it gives a symplectic structure on $\mathcal{P}^D$.

Remark 5.10. In the case $D = (\varpi)$, the form ω is still nondegenerate, but the contribution of the points P_m vanishes. This corresponds to the case of Riemann surfaces without marked points in the Hitchin approach.

5.6 Hamiltonian theory

Following the lines of [33] we show here that the hierarchies in Theorem 5.6 are Hamiltonian for all classical Lie algebras in question, and we compute the corresponding Hamiltonians.

For a vector field e on $\mathcal{L}^D$, let $i_e\omega$ be the 1-form defined by $i_e\omega(X) = \omega(e, X)$ (where X is an arbitrary vector field). By definition, a vector field ∂_t is Hamiltonian if $i_{\partial_t}\omega = \delta H$, where H is some function called the Hamiltonian of ∂_t. The following theorem is the main purpose of this section.

Theorem 5.11. *Let ∂_a be a vector field defined by* (5.4.4). *Then*

$$i_{\partial_a}\omega = \delta H_a,$$

where

$$H_a = -\frac{1}{k+1}\operatorname{res}_{P_i}\operatorname{tr}\bigl(w^{-m}L^{k+1}\bigr)\varpi,\ a = (P_i, k, m).$$

Before we begin with the proof of this theorem, let us consider some properties of L, $\partial_a + M_a$ and their spectra. First, these operators commute due to the Lax equation, hence the rows of Ψ are eigenvectors for both of them. The diagonal forms of those two operators are defined as

$$K = \Psi L \Psi^{-1},\ F_a = \Psi(\partial_a + M_a)\Psi^{-1}. \tag{5.6.1}$$

Equivalently

$$\Psi L = K\Psi,\ \partial_a \Psi = \Psi M_a - F_a \Psi. \tag{5.6.2}$$

These relations imply

$$\partial_a K = 0. \tag{5.6.3}$$

As shown in Section 5.5, L is conjugated to a function holomorphic at the γ-points; hence its spectrum K is also holomorphic there. The same is true for F_a (where a is arbitrary). To prove this we need more information, which is provided by Lemma 5.12 below.

The elements of K, considered as functions of $P \in \Sigma$, certainly correspond to the branches of the function $\varkappa(z)$ given by the spectral equation $\det(L(z) - \varkappa) = 0$. Let us also regard the eigenvectors of L as functions of P.

Let $\mathcal{F}$ be the space of meromorphic vector-valued functions ψ on Σ holomorphic except at $P_1, \dots, P_N$ and γ's, such that

$$\psi(z) = \nu\frac{\alpha}{z} + \psi_0 + \cdots$$

at any point γ (z is a local coordinate centered at γ). This space already occured in Section 2.2.

Lemma 5.12. *The eigenvectors of L belong to the space $\mathcal{F}$.*

Proof. It is instructive to begin with an example of $\mathfrak{g} = \mathfrak{sl}(2)$. Making use of gauge freedom (like in Section 5.5), let us assume that $\alpha = (1,0)^t$. Then $L_0\alpha = \varkappa\alpha$ implies

$$L_0 = \begin{pmatrix} a & b \\ 0 & -a \end{pmatrix},$$

where $a, b \in \mathbb{C}$. Let $\beta = (\beta_1, \beta_2)^t$, then by $\beta^t\alpha = 0$ we have $\beta_1 = 0$.

Thus, in the neighborhood of γ we have

$$\begin{aligned} L = \frac{\alpha\beta^t}{z} + L_0 + O(z) &= \frac{1}{z}\begin{pmatrix} 0 & \beta_2 \\ 0 & 0 \end{pmatrix} + \begin{pmatrix} a & b \\ 0 & -a \end{pmatrix} + O(z) \\ &= \begin{pmatrix} a & b + \frac{\beta_2}{z} \\ O(z) & -a \end{pmatrix} + F(z), \end{aligned} \tag{5.6.4}$$

where $F(z)$ is a 2×2-matrix, $F(z) = O(z)$, $F_{21}(z) = O(z^2)$. The characteristic equation of L with the spectral parameter λ has the form

$$\lambda^2 - a^2 - \tilde{\beta}_2 + O(z) = 0,$$

where $\tilde{\beta}_2 \in \mathbb{C}$. Therefore

$$\lambda = \pm\sqrt{a^2 + \tilde{\beta}_2 + O(z)}$$

is holomorphic for $z \sim 0$. By this we have shown the holomorphy of the spectra in this example. Now take a certain λ and consider the corresponding eigenvector $(x_1, x_2)^t$. By definition, it satisfies the linear equations

$$\begin{cases} (a-\lambda)x_1 + \left(b + \dfrac{\beta_2}{z}\right) x_2 = O(z) \\ zx_1 - (a+\lambda)x_2 = O(z) \end{cases}$$

which are compatible. We have from the second equation

$$x_1 = \frac{(a+\lambda)}{z} x_2 + O(1),$$

hence

$$\begin{pmatrix} x_1 \\ x_2 \end{pmatrix} = \frac{(a+\lambda)x_2}{z}\alpha + \begin{pmatrix} O(1) \\ x_2 \end{pmatrix},$$

i.e., it has a required form.

There is no essential difference in the general case. The main argument holds true, namely that all entries of the first column of L are of the order 1 in the variable z, except perhaps for the first one (see e.g. the proof of Lemma 5.7 in the symplectic case). For this reason, the first coordinate of an eigenvector has the order $O(1/z)$, and hence we have

$$x = \nu\frac{\alpha}{z} + O(1), \quad \nu \in \mathbb{C}$$

for this vector. □

The columns of Ψ^{-1} are right eigenvectors of the matrix L. By Lemma 5.12 we obtain that all columns of the matrix $\operatorname{res}_\gamma \Psi^{-1}$ are multiples of α. Hence there exist $\widehat{\beta}^t \in \mathbb{C}^n$ such that $\operatorname{res}_\gamma \Psi^{-1} = \alpha\widehat{\beta}^t$. It is convenient to represent $\widehat{\beta}$ in the form $\widehat{\beta} = \tilde{\beta}\sigma$ where $\tilde{\beta} \in \mathbb{C}^n$. Then by involution $\operatorname{res}_\gamma \Psi = -\varepsilon\tilde{\beta}\alpha^t\sigma$. In the cases $\mathfrak{g} = \mathfrak{gl}(n)$, $\mathfrak{g} = \mathfrak{sl}(n)$, the last residue vanishes. Thus we obtain the following expansions in the neighborhood of γ:

$$\begin{aligned} \Psi(z) &= \frac{\Psi_{-1}}{z - z_\gamma} + \Psi_0 + \Psi_1(z - z_\gamma) + \cdots, \quad \Psi_{-1} = -\varepsilon\tilde{\beta}\alpha^t\sigma, \\ \Psi^{-1}(z) &= \frac{\tilde{\Psi}_{-1}}{z - z_\gamma} + \tilde{\Psi}_0 + \tilde{\Psi}_1(z - z_\gamma) + \cdots, \quad \tilde{\Psi}_{-1} = \alpha\tilde{\beta}^t\sigma \end{aligned} \tag{5.6.5}$$

where ε, σ are as usual.

Lemma 5.13. *The following relations hold true:*

$$\Psi_0\alpha = 0, \quad \varepsilon\alpha^t\sigma\tilde{\Psi}_0 = 0 \tag{5.6.6}$$

except perhaps in the following cases:

(1) $\mathfrak{g} = \mathfrak{gl}(n), \mathfrak{sl}(n)$, $\beta = 0$;

(2) $\mathfrak{g} = \mathfrak{so}(n)$, $\beta \in \mathbb{C}\alpha$;

(3) $\mathfrak{g} = \mathfrak{sp}(2n)$, $\nu = 0$, *and* $\beta \in \mathbb{C}\alpha$.

In addition we have the obvious relations

$$\Psi_{-1}\alpha = 0, \quad \alpha^t\sigma\tilde{\Psi}_{-1} = 0, \quad \Psi_{-1}\beta = 0, \quad \beta^t\sigma\tilde{\Psi}_{-1} = 0. \tag{5.6.7}$$

Proof of Lemma 5.13. To prove that $\Psi_0\alpha = 0$ we address the relation $\Psi L = K\Psi$.

First, consider the cases $\mathfrak{g} = \mathfrak{gl}(n), \mathfrak{sl}(n)$. Then $\varepsilon = 0$, and Ψ is holomorphic at γ. Since K is holomorphic there also, we conclude that

$$\operatorname{res}_\gamma \Psi L = \Psi_0\alpha\beta^t = 0.$$

If $\beta \neq 0$, then $\Psi_0\alpha = 0$.

Next, let $\mathfrak{g} = \mathfrak{sp}(2n)$. The same relation $\Psi L = K\Psi$ now gives

$$\left(\frac{\Psi_{-1}}{z} + \Psi_0 + \Psi_1 z + \cdots\right)\left(\nu\frac{\alpha\alpha^t\sigma}{z^2} + \cdots\right) = O\left(\frac{1}{z}\right).$$

In the term with z^{-2} we obtain $\nu\Psi_0\alpha\alpha^t\sigma = 0$ (by (5.6.7) Ψ_{-1} does not affect this relation). Since $\alpha \neq 0$ and σ is a nondegenerate quadratic form, we have $\nu\Psi_0\alpha = 0$. Consider now the same relation in the form $L\Psi^{-1} = \Psi^{-1}K$. For the left-hand side we obtain by computation $L\Psi^{-1} = \nu\alpha\alpha^t\sigma\tilde{\Psi}_0 z^{-2} + O(z^{-1})$ while the right-hand side has at most simple pole. As above, we conclude that $\nu\alpha^t\sigma\tilde{\Psi}_0 = 0$. If $\nu \neq 0$, then the relations (5.6.6) are proven.

Let us now consider the cases $\mathfrak{g} = \mathfrak{so}(n)$, and $\mathfrak{g} = \mathfrak{sp}(2n)$, $\nu = 0$, together. Setting equal the residues in the both hand sides of the relation $\Psi L = K\Psi$ we obtain ($\varepsilon = -1$, $\sigma = id$)

$$-\varepsilon\tilde{\beta}\alpha^t\sigma L_0 + \Psi_0(\alpha\beta^t + \varepsilon\beta\alpha^t)\sigma = -\varepsilon K_0\tilde{\beta}\alpha^t\sigma.$$

If $\beta \notin \mathbb{C}\alpha$, then there exists a $\beta' \in \mathbb{C}^n$ such that $\alpha^t\sigma\beta' = 0$, $\beta^t\sigma\beta' \neq 0$. Observe also that $\alpha^t\sigma L_0 = \varkappa\alpha^t\sigma$. Multiplying the above relation by β', we obtain $\Psi_0\alpha = 0$, and by the involution $\alpha^t\sigma\tilde{\Psi}_0 = 0$. □

For the remainder of this section we assume that the exceptional cases of Lemma 5.13 do not take place.

Lemma 5.14. *The matrix-valued functions K and F_a are holomorphic at all γ-points provided* (5.6.6) *hold there.*

Proof. As it was mentioned on page 118, the holomorphy of K follows from the results of Section 5.5. Again we want to stress that in the proof of Lemma 5.7 it is shown that L is conjugated to a function holomorphic at the γ-points, and hence its spectrum K is also holomorphic there.

Let us now prove the holomorphy of the spectrum of $\partial_a + M_a$. Indeed, this spectrum is expressed by the matrix $F_a = \Psi(\partial_a + M_a)\Psi^{-1}$. Using the expansions for M_a and Ψ^{-1} at a γ, we obtain

$$\begin{aligned}
&(\partial_a + M_a)\Psi^{-1} = \\
&\quad \frac{(\lambda\alpha\alpha^t\sigma)(\alpha\tilde{\beta}^t)}{(z-z_\gamma)^3} + \frac{(\partial_a z_\gamma)\alpha\tilde{\beta}^t + \lambda\alpha\alpha^t\sigma\tilde{\Psi}_0 + (\alpha\mu_a^t + \varepsilon\mu_a\alpha^t)\sigma\alpha\tilde{\beta}^t}{(z-z_\gamma)^2} \\
&\quad + \frac{(\partial_a\alpha)\tilde{\beta}^t + \alpha(\partial_a\tilde{\beta}^t) + \lambda\alpha\alpha^t\sigma\tilde{\Psi}_1 + (\alpha\mu_a^t + \varepsilon\mu_a\alpha^t)\sigma\tilde{\Psi}_0 + M_{0a}\alpha\tilde{\beta}^t}{z-z_\gamma} \\
&\quad + (\partial_a\tilde{\Psi}_0 + M_{a,0}\tilde{\Psi}_0) - (\partial_a z_\gamma)\tilde{\Psi}_1 + (\alpha\mu_a^t + \varepsilon\mu_a\alpha^t)\sigma\tilde{\Psi}_1 \\
&\quad + \lambda_a\alpha\alpha^t\sigma\tilde{\Psi}_2 + M_{a,1}\tilde{\Psi}_{-1} + O(z-z_\gamma).
\end{aligned}$$

After omitting the terms vanishing by the relations (5.6.6), and the relations given by definition of a Lax operator algebra, we obtain

$$\begin{aligned}
(\partial_a + M_a)\Psi^{-1} &= \frac{(\partial_a z_\gamma + \mu_a^t\sigma\alpha)\alpha\tilde{\beta}^t}{(z-z_\gamma)^2} + \frac{(\partial_a\alpha + M_{a,0}\alpha)\tilde{\beta}^t}{z-z_\gamma} + \frac{\alpha\hat{\beta}^t}{z-z_\gamma} \\
&\quad + (\partial_a\tilde{\Psi}_0 + M_{a,0}\tilde{\Psi}_0) - (\partial_a z_\gamma)\tilde{\Psi}_1 + (\alpha\mu_a^t + \varepsilon\mu_a\alpha^t)\sigma\tilde{\Psi}_1 \\
&\quad + \lambda_a\alpha\alpha^t\sigma\tilde{\Psi}_2 + M_{a,1}\tilde{\Psi}_{-1} + O(z-z_\gamma)
\end{aligned}$$

where $\hat{\beta}^t = \partial_a\tilde{\beta}^t + \lambda\alpha^t\sigma\tilde{\Psi}_1 + \mu_a^t\sigma\Psi_0$.

Due to (5.3.3), the first summand vanishes and the second summand is equal to $\varkappa\alpha\tilde{\beta}^t$. Hence

$$\begin{aligned}
(\partial_a + M_a)\Psi^{-1} &= \frac{\alpha\hat{\beta}^t + \varkappa\alpha\tilde{\beta}^t}{z-z_\gamma} + \lambda_a\alpha\alpha^t\sigma\tilde{\Psi}_2 + M_{a,1}\tilde{\Psi}_{-1} + (\partial_a\tilde{\Psi}_0 + M_{a,0}\tilde{\Psi}_0) \\
&\quad - (\partial_a z_\gamma)\tilde{\Psi}_1 + (\alpha\mu_a^t + \varepsilon\mu_a\alpha^t)\sigma\tilde{\Psi}_1 + O(z-z_\gamma).
\end{aligned}$$

Continuing on, we have

$$\begin{aligned}
&\Psi(\partial_a + M_a)\Psi^{-1} = \\
&\quad \left(\frac{\Psi_{-1}}{z-z_\gamma} + \Psi_0\right)\left(\frac{\alpha\hat{\beta}^t + \varkappa\alpha\tilde{\beta}^t}{z-z_\gamma} + \lambda_a\alpha\alpha^t\sigma\tilde{\Psi}_2 + M_{a,1}\tilde{\Psi}_{-1}\right) \\
&\quad + \frac{\Psi_{-1}}{z-z_\gamma}\left((\partial_a\tilde{\Psi}_0 + M_{a,0}\tilde{\Psi}_0) - (\partial_a z_\gamma)\tilde{\Psi}_1 + (\alpha\mu_a^t + \varepsilon\mu_a\alpha^t)\sigma\tilde{\Psi}_1\right) + O(1).
\end{aligned}$$

The first line of the last relation vanishes by $\Psi_{-1}\alpha = 0$ (5.6.7), $\Psi_0\alpha = 0$ (5.6.6), and (5.1.6) (since $\Psi_{-1}M_{a,1}\tilde{\Psi}_{-1} = -\varepsilon\tilde{\beta}(\alpha^t\sigma M_{a,1}\alpha)\tilde{\beta}\sigma = 0$).

Let us consider the second line term by term. First, by (5.3.3)

$$\varepsilon\alpha^t\sigma(-\partial_a z_\gamma) = \varepsilon\alpha^t\sigma\mu_a^t\sigma\alpha$$

($\varepsilon\alpha^t\sigma$ comes from the common factor Ψ_{-1}). Then

$$\varepsilon\alpha^t\sigma(\alpha\mu_a^t + \varepsilon\mu_a\alpha^t)\sigma = \alpha^t\sigma\mu_a\alpha^t\sigma.$$

Observe that $\alpha^t\sigma\mu_a = -\varepsilon\mu_a^t\sigma\alpha$ (and it is a scalar). Hence the sum of two terms containing $\tilde{\Psi}_1$ in the last expression for $\Psi(\partial_a + M_a)\Psi^{-1}$ vanishes.

Finally, let us prove that

$$\varepsilon\alpha^t\sigma(\partial_a\tilde{\Psi}_0 + M_{a,0}\tilde{\Psi}_0) = 0.$$

Assume that $\varepsilon \neq 0$ (otherwise it is obvious). Then $\alpha^t\sigma\tilde{\Psi}_0 = 0$, and by derivation of this relation we obtain $(\partial_a\alpha^t)\sigma\tilde{\Psi}_0 + \alpha^t\sigma(\partial_a\tilde{\Psi}_0) = 0$. By (5.3.3) we also have $\partial_a\alpha^t\sigma = \alpha^t\sigma M_{a,0} + \varkappa\alpha^t\sigma$, and hence

$$\alpha^t\sigma(\partial_a\tilde{\Psi}_0) = -(\partial_a\alpha^t)\sigma\tilde{\Psi}_0 = -(\alpha^t\sigma M_{a,0} + \varkappa\alpha^t\sigma)\tilde{\Psi}_0.$$

Continuing on, we obtain

$$\alpha^t\sigma(\partial_a\tilde{\Psi}_0 + M_{a,0}\tilde{\Psi}_0) = -\varkappa\alpha^t\sigma\tilde{\Psi}_0 = 0.$$

Hence

$$\Psi(\partial_a + M_a)\Psi^{-1} = O(1).$$ □

Proof of Theorem 5.11. Modulo Lemma 5.14, the proof of the theorem is the same as in [33]. We give it here for completeness.

By definition,

$$i_{\partial_a}\omega = \omega(\partial_a,\cdot) = -\frac{1}{2}\Big(\sum_{s=1}^{K}\operatorname{res}_{\gamma_s}\Lambda + \sum_{i=1}^{N}\operatorname{res}_{P_i}\Lambda\Big),$$

where $\Lambda = \Omega(\partial_a,\cdot)$. By $\delta\Psi(\partial_a) = \partial_a\Psi$ and $\delta L(\partial_a) = \partial_a L$ (the evaluation of a differential on a vector field is the derivative along that vector field), we have

$$\Lambda = \operatorname{tr}\big(\partial_a\Psi\cdot\delta L\cdot\Psi^{-1} - \delta\Psi\cdot\partial_a L\cdot\Psi^{-1} - \partial_a K\cdot\delta\Psi\cdot\Psi^{-1} + \delta K\cdot\partial_a\Psi\cdot\Psi^{-1}\big).$$

By (5.6.2) and the Lax equation

$$\begin{aligned}\Lambda &= \operatorname{tr}\big((\Psi M_a - F_a\Psi)\delta L\cdot\Psi^{-1} - \delta\Psi[L, M_a]\Psi^{-1} + \delta K(\Psi M_a - F_a\Psi)\Psi^{-1}\big)\\ &= \operatorname{tr}\big(M_a\delta L - F_a\Psi\delta L\cdot\Psi^{-1} - \delta\Psi[L, M_a]\Psi^{-1} + \delta K\Psi M_a\Psi^{-1} - \delta K F_a\big).\end{aligned}$$

Let us transform the middle term. By $\Psi L = K\Psi$ we have $\delta\Psi \cdot L = -\Psi\delta L + \delta K\Psi + K\delta\Psi$. Hence

$$\begin{aligned}
&\operatorname{tr}\delta\Psi[L, M_a]\Psi^{-1}\\
&\quad= \operatorname{tr}\big((\delta\Psi \cdot L)M_a\Psi^{-1} - \delta\Psi M_a L\Psi^{-1}\big)\\
&\quad= \operatorname{tr}\big((-\Psi\delta L + \delta K\Psi + K\delta\Psi)M_a\Psi^{-1} - \delta\Psi M_a L\Psi^{-1}\big)\\
&\quad= \operatorname{tr}\big(-\Psi\delta L M_a\Psi^{-1} + \delta K\Psi M_a\Psi^{-1} + K\delta\Psi M_a\Psi^{-1} - \delta\Psi M_a L\Psi^{-1}\big).
\end{aligned}$$

The last two terms annihilate because

$$\operatorname{tr}\big(\delta\Psi M_a L\Psi^{-1}\big) = \operatorname{tr}\big(\delta\Psi M_a\Psi^{-1}(\Psi L\Psi^{-1})\big) = \operatorname{tr}\big(\delta\Psi M_a\Psi^{-1}K\big),$$

and we obtain

$$\operatorname{tr}\delta\Psi[L, M_a]\Psi^{-1} = \operatorname{tr}\big(-\delta L M_a + \delta K\Psi M_a\Psi^{-1}\big).$$

Substituting that to the last expression for Λ, we obtain

$$\Lambda = \operatorname{tr}\big(2M_a\delta L - F_a\Psi\delta L \cdot \Psi^{-1} - \delta K F_a\big).$$

The last two terms are equal (under the symbol of tr) and can be obtained by replacing $\Psi\delta L$ with $-\delta\Psi L + \delta K\Psi + K\delta\Psi$. Hence we finally obtain

$$\Lambda = \operatorname{tr}(2M_a\delta L - 2\delta K F_a).$$

This gives

$$i_{\partial_a}\omega = \sum_{j=1}^{N} \operatorname{res}_{P_j} \operatorname{tr}(\delta K\, F_a)\varpi - R_a, \tag{5.6.8}$$

where

$$R_a = \sum_{s=1}^{K} \operatorname{res}_{\gamma_s} \operatorname{tr}(\delta L M_a)\varpi + \sum_{j=1}^{N} \operatorname{res}_{P_j} \operatorname{tr}(\delta L M_a)\varpi. \tag{5.6.9}$$

A priori there should be a sum of residues of $\operatorname{tr}(\delta K\, F_a)\varpi$ over the γ-points in (5.6.8) but it vanishes because δK and F_a are holomorphic at those points (Lemma 5.14). Observe also that the function F_a has singularities other than γ-points and the points P_j. Indeed, $F_a = -\partial_a\Psi \cdot \Psi^{-1} - \Psi M_a\Psi^{-1}$, and Ψ^{-1} has poles at the branching points of eigenvalues of L.

Contrary to this, L and M_a are holomorphic everywhere except at the points γ_s and P_j. Hence $R_a = 0$ as the sum of residues of a meromorphic 1-form over all its poles. Moreover, by the construction of M_a, the matrix F_a is holomorphic at P_j, $j \neq i$ for all times a corresponding to the point P_i. For this reason, for such times

$$i_{\partial_a}\omega = \operatorname{res}_{P_i} \operatorname{tr}(\delta K\, F_a)\varpi.$$

But $F_a = -\partial_a\Psi \cdot \Psi^{-1} - \Psi M_a \Psi^{-1}$, and $\partial_a\Psi \cdot \Psi^{-1}$ is holomorphic at P_i because the coordinate of P_i is independent of any time; therefore, $\partial_a\Psi$ and Ψ have the same order at P_i. Hence $F_a = -\Psi M_a \Psi^{-1} + O(1)$. By definition, $M_a = w^{-m}L^k + O(1)$ at P_i for $a = (P_i, k, m)$. This implies $\Psi M_a \Psi^{-1} = w^{-m}\Psi L^k \Psi^{-1} + O(1)$, since Ψ is holomorphically invertible at P_i. However, $\Psi L^k \Psi^{-1} = K^k$, and hence $F_a = w^{-m}K^k + O(1)$ at P_i. Since $\delta K \varpi$ is holomorphic at P_i, we obtain

$$\begin{aligned} i_{\partial_a}\omega &= -\operatorname{res}_{P_i} \operatorname{tr}\big(w_i^{-m}\delta K\, K^k\big)\varpi = -\frac{1}{k+1}\operatorname{res}_{P_i} \delta\operatorname{tr}\big(w_i^{-m}K^{k+1}\big) \\ &= -\frac{1}{k+1}\delta \operatorname{res}_{P_i} \operatorname{tr}\big(w_i^{-m}L^{k+1}\big) = \delta H_a. \end{aligned}$$

□

The Hamiltonians H_a are in involution, since they depend only on the spectral parameters. We refer to [33] for more details.

5.7 Examples: Calogero–Moser systems

Let us begin with the example considered in [33] – the elliptic Calogero–Moser model corresponding to the root system A_n. Define a $\mathfrak{gl}(n)$-valued Lax operator by

$$L_{ij} = f_{ij}\frac{\sigma(z+q_j-q_i)\sigma(z-q_j)\sigma(q_i)}{\sigma(z)\sigma(z-q_i)\sigma(q_j-q_i)\sigma(q_j)} \ (i \neq j), \quad L_{jj} = p_j, \tag{5.7.1}$$

where σ (and $\wp$ below) are Weierstrass functions, $f_{ij} \in \mathbb{C}$ are constant. Up to the constants f_{ij} this form of L is determined by the requirements that L is elliptic and that it has simple poles at the points $z = q_i$ $(i = 1, \dots, n)$ and $z = 0$. The last is the only immovable pole. By reduction of the remaining gauge freedom, it is obtained in [33] that $f_{ij}f_{ji} = 1$. For the second order Hamiltonian corresponding to that pole, we have, according to Theorem 5.11, up to normalization,

$$H = \operatorname{res}_{z=0}\ z^{-1}\Big(-\frac{1}{2}\sum_{j=1}^{n} p_j^2 - \sum_{i<j} L_{ij}L_{ji}\Big).$$

By the addition theorem for Weierstrass functions

$$-L_{ij}L_{ji} = \frac{\sigma(z+q_i-q_j)\sigma(z+q_j-q_i)}{\sigma(z)^2\sigma(q_i-q_j)^2} = \wp(q_i-q_j) - \wp(z),$$

hence

$$H = -\frac{1}{2}\sum_{j=1}^{n} p_j^2 + \sum_{i<j}\wp(q_i-q_j).$$

Let us now consider the cases $\mathfrak{g} = \mathfrak{so}(2n)$, $\mathfrak{g} = \mathfrak{sp}(2n)$. In the first case we assume that the Tyurin divisor is supported at $K = 2n$ points denoted by $\pm q_1, \dots, \pm q_n$. In the second case, $K = 2(n+1)$, and $\pm q_1, \ldots, \pm q_{n+1}$ are the corresponding points. As earlier, we denote the space of the M-operators corresponding to a positive divisor D (of fixed points) by $\mathcal{N}^D$. The next lemma is analogous to Lemma 5.5.

Lemma 5.15. *For* $\mathfrak{g} = \mathfrak{so}(2n), \mathfrak{sp}(2n)$

$$\dim \mathcal{N}^D = (\dim \mathfrak{g}^\diamond)(\deg D + 2).$$

Proof. We keep here the same line of argument as in the proof of Lemma 5.5, the only difference being that l is twice as big (see page 111 for the definition of l). Namely, $l = 2n$ for $\mathfrak{g} = \mathfrak{so}(2n)$, and $l = 2(n+1)$ for $\mathfrak{g} = \mathfrak{sp}(2n)$.

For this reason, in place of (5.4.2) we obtain

$$2 \dim \mathfrak{g}^\diamond = (r - r_{\mathfrak{sp}})l \tag{5.7.2}$$

which gives an extra contribution to $\dim \mathcal{N}^D$ equal to $\dim \mathfrak{g}^\diamond$. This proves the Lemma. □

In order for the other results of Section 5.4 to hold true, we replace the normalization condition (iii) in the definition of M_a (page 112) with similar conditions at two points.

Next, we want to obtain the Calogero–Moser system corresponding to D_n by means of our approach. For this purpose, let us take the Lax operator in the form

$$L = \begin{pmatrix} A & \boxed{\begin{matrix}0\\ \vdots\\ 0\end{matrix}} & B \\ \boxed{0\ldots\ldots 0} & 0 & \boxed{0\ldots\ldots 0} \\ C & \boxed{\begin{matrix}0\\ \vdots\\ 0\end{matrix}} & -A^t \end{pmatrix}, \tag{5.7.3}$$

where A, B, C are $n \times n$-matrices, $B = -B^t$, $C = -C^t$. It is a standard form of an element of $\mathfrak{so}(2n)$ embedded in $\mathfrak{so}(2n+1)$. Observe that the "null-cross" formed by the $(n+1)$-th row and $(n+1)$-th column is not necessarily trivial for the corresponding M-operators.

Assume A to be given by (5.7.1). For $i < j$ take

$$B_{ij} = f_{ij}^B \frac{\sigma(z + q_j + q_i)\sigma(z - q_j)}{\sigma(z)\sigma(z + q_i)\sigma(q_i + q_j)}, \quad C_{ji} = f_{ji}^C \frac{\sigma(z - q_j - q_i)\sigma(z + q_i)}{\sigma(z)\sigma(z - q_j)\sigma(q_i + q_j)}, \tag{5.7.4}$$

where $f_{ij}^B, f_{ij}^C \in \mathbb{C}$ are constant. These relations determine the matrices B and C due to skew symmetry. Similar to the case of $\mathfrak{gl}(n)$ we obtain $f_{ij}^B f_{ji}^C = -1$ by reduction of the remaining gauge freedom (taking into account the relation $\alpha^t \sigma \alpha = 0$ which in this case descends to $\alpha''^t\alpha'' = 0$). For the Hamiltonian we have

$$H = -\operatorname{res}_{z=0}\ z^{-1}\Big(\sum_{i=1}^{n} p_i^2 + 2\sum_{i<j} A_{ij}A_{ji} + 2\sum_{i<j} B_{ij}C_{ji}\Big)$$
$$= -\sum_{i=1}^{n} p_i^2 + 2\sum_{i<j}\wp(q_i - q_j) + 2\sum_{i<j}\wp(q_i + q_j)$$

which is a conventional form of the second-order Hamiltonian of the elliptic D_n Calogero–Moser model.

Remark 5.16. There is a simple relation between our Lax operator and the one given in [13]. In the case $\mathfrak{g} = \mathfrak{so}(n)$ the latter is of the form

$$\widetilde{L} = \begin{pmatrix} P & 0 \\ 0 & -P \end{pmatrix} + \begin{pmatrix} A & B \\ C & A^t \end{pmatrix}, \quad B^t = B, \quad C^t = C, \tag{5.7.5}$$

where $P = \operatorname{diag}(p_1, \dots, p_n)$, A, B, and C are matrices with zero diagonals, and the rest of the entries are Baker-Akhieser functions on the elliptic curve. In these terms our Lax operator is of the form

$$L = \begin{pmatrix} P & 0 \\ 0 & -P \end{pmatrix} + \begin{pmatrix} A & B \\ C & -A^t \end{pmatrix}, \quad B^t = -B, \quad C^t = -C. \tag{5.7.6}$$

The following transformation takes L to $\widetilde{L}$:

$$P \to P, \quad A \to A, \quad B = b - b^t \to B' = b + b^t, \quad C = c - c^t \to C' = -(c + c^t)$$

where b and c are upper triangular matrices. For example, for $\mathfrak{g} = so(4)$ it appears as follows:

$$\begin{pmatrix} p_1 & a_{12} & 0 & b \\ a_{21} & p_2 & -b & 0 \\ 0 & c & -p_1 & -a_{21} \\ -c & 0 & -a_{12} & -p_2 \end{pmatrix} \to \begin{pmatrix} p_1 & a_{12} & 0 & b \\ a_{21} & p_2 & b & 0 \\ 0 & -c & -p_1 & a_{21} \\ -c & 0 & a_{12} & -p_2 \end{pmatrix}, \tag{5.7.7}$$

where $b, c \in \mathbb{C}$. The transformation preserves $\operatorname{tr} L^2$, $\operatorname{tr} L^3 = 0$ and $\operatorname{tr} L^4$.

If the entries of L are given by (5.7.4), (5.7.1), then the composition of that transformation and the conjugation by the diagonal matrix $e^{q\zeta(z)}$ (where ζ is the Weierstrass ζ-function, $q = \operatorname{diag}(q_1, \dots, q_n, -q_1, \dots, -q_n)$) results in the Lax operator [13].

There are similar relations between the two Lax operators for other classical Lie algebras $\mathfrak{g}$. For $\mathfrak{g} = \mathfrak{gl}(n)$ the gauge equivalence of L with the known Lax operator [24] is mentioned in [33].

Looking at the Calogero–Moser systems corresponding to the dual root systems B_n and C_n, let us note that each of them can be obtained using any of Lie algebras $\mathfrak{g} = \mathfrak{sp}(2n)$, $\mathfrak{g} = \mathfrak{so}(2n+1)$ within our approach.

Let us start with the C_n Calogero–Moser system using $\mathfrak{g} = \mathfrak{sp}(2n)$ for this purpose. Define $\mathcal{N}^D$ as the space of M-operators taking values in $\mathfrak{g}^{\diamond} = \mathfrak{tsp}(2n)$, with $n+1$ pairs of double poles $\pm q_i$ ($i = 1, \dots, n+1$). Thus,

$$M = \begin{pmatrix} 0 & -a^t \; b^t & c \\ 0 & \boxed{\begin{matrix} A & B \\ C & -A^t \end{matrix}} & \begin{matrix} b \\ a \end{matrix} \\ 0 & 0 & 0 \end{pmatrix},$$

where $a, b \in \mathbb{C}^n$, $c \in \mathbb{C}$, A, B, C are $n \times n$-matrices, $B = B^t$, $C = C^t$.

Let us take L in the same form as M where the corresponding elements A_{ij} are defined by the right-hand sides of (5.7.1), B_{ij} and C_{ij} are defined by (5.7.4). The relations (5.7.4) make sense for $i = j$ as well, and in this case their contribution to the second-order Hamiltonian is equal to

$$B_{ii} C_{ii} = f_{ii}^B f_{ii}^C \big(\wp(2q_i) - \wp(z)\big)$$

where we can set $f_{ii}^B f_{ii}^C = 2$. The submatrices a and b of the matrix L do not contribute to the Hamiltonians regardless of any explicit form they might have, and the constant (in p, q-variables) term $\wp(z)$ is negligible. Hence

$$H = -\sum_{i=1}^{n} p_i^2 + 2\sum_{i<j} \wp(q_i - q_j) + 2\sum_{i<j} \wp(q_i + q_j) + 2\sum_{i=1}^{n} \wp(2q_i),$$

which is a conventional form of the second-order Hamiltonian of the elliptic Calogero–Moser model in the symplectic case.

Now obtain the B_n Calogero–Moser system using $\mathfrak{g} = \mathfrak{sp}(2n)$. For this we only need to replace the diagonal elements of the matrices B and C in the previous consideration. Let us take

$$B_{ii} = f_{ii}^B \frac{\sigma(z+q_i)\sigma(z-\frac{1}{2}q_i)}{\sigma(z)\sigma(z+\frac{1}{2}q_i)\sigma(q_i)}, \quad C_{ii} = f_{ii}^C \frac{\sigma(z-q_i)\sigma(z+\frac{1}{2}q_i)}{\sigma(z)\sigma(z-\frac{1}{2}q_i)\sigma(q_i)}. \tag{5.7.8}$$

These functions are well defined on the 4-sheeted covering of our elliptic curve corresponding to the torus with redoubled periods. The corresponding contribution to the Hamiltonian is equal to

$$\frac{\sigma(z+q_i)\sigma(z-q_i)}{\sigma(z)^2\sigma(q_i)^2} = \wp(q_i) - \wp(z),$$

and is well defined on the original elliptic curve. Hence

$$H = -\sum_{i=1}^{n} p_i^2 + 2\sum_{i<j} \wp(q_i - q_j) + 2\sum_{i<j} \wp(q_i + q_j) + 2\sum_{i=1}^{n} \wp(q_i),$$

which is a conventional form of the second-order Hamiltonian of the elliptic B_n Calogero–Moser model.

Observe that we use L- and M-operators with $4n$ moving simple poles in this approach. Respectively, we have

$$\dim \mathcal{N}^D = (\dim \mathfrak{g}^\diamond)(\deg D + 4)$$

instead the corresponding formulae of Lemmas 5.15 and 5.5. Hence, in order for the results of Section 5.4 to hold true, we must normalize at four points in the definition of the operator M_a ((iii), page 112).

Both B_n and C_n Calogero–Moser systems can also be obtained using the $\mathfrak{so}(2n+1)$-valued Lax operators, as was mentioned above. For this purpose we must use the Lax operators of the form (5.7.3), where zero entries in the "null-cross" must be replaced with what we put on the diagonal of the matrices B, C in the corresponding cases.

Chapter 6

Lax integrable systems and conformal field theory

This chapter brings together the material from all the previous chapters.

We address here the following problem: given a Lax integrable system of the type discussed above, construct a unitary projective representation of the corresponding Lie algebra of Hamiltonian vector fields. For the Lax equations in question, we propose a way to represent Hamiltonian vector fields by covariant derivatives with respect to the Knizhnik–Zamolodchikov connection. This is a Dirac-type prequantization from the physical point of view. The idea of quantization of Hitchin systems by means the Knizhnik–Zamolodchikov connection was addressed, or at least mentioned, many times in the theoretical physics literature (D. Ivanov [14], G. Felder and Ch. Wieczerkowski [7], M. A. Olshanetsky and A. M. Levin [36,37]), but only second-order Hamiltonians have been dealt with. Unexpectedly, we observed such a relation for all Hamiltonians, and, moreover, for all observables of the Hamiltonian system given by the Lax equations in question [72].

In Section 6.1 we briefly outline the main ideas and results of the chapter. Details are given in Sections 6.2–6.7. In particular, in Sections 6.2 and 6.3 the maximal Lie subalgebra in $\mathcal{L}$ commuting with a generic $L \in \mathcal{L}$ is described in terms of the spectral curve of L. Sections 6.8 and 6.9 are devoted to a discussion. In particular the relation to the Seiberg–Witten theory is briefly discussed.

6.1 Conformal field theory related to a Lax integrable system

By conformal field theory we mean a family of Riemann surfaces, a finite rank bundle (of *conformal blocks*) on this family, and a flat connection (*Knizhnik–Zamolodchikov connection*) on this bundle. In Chapter 3 we considered a moduli space of Riemann surfaces with marked points as such a family. Here, however, we consider the universal bundle of spectral curves over the phase space $\mathcal{P}^D$ of the integrable system (Figure 6.1) defined in Section 5.5.

For every $L \in \mathcal{P}^D$ (all arguments of L are fixed except for z) the curve Σ_L given by the equation $\det(L(z) - \lambda) = 0$ is called a *spectral curve* of L. It is a n-fold branch covering of Σ.

Thus we obtained a family of Riemann surfaces over $\mathcal{P}^D$. Following the lines of Chapter 3, we construct an analog of the Knizhnik–Zamolodchikov connection for this family. To this end, we construct a sheaf of admissible representations of the Lax

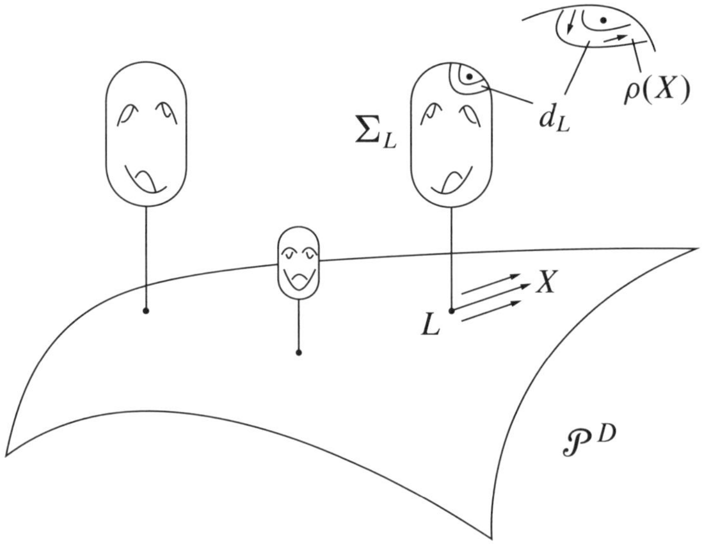

Figure 6.1. Family of spectral curves over the phase space.

operator algebra canonically related to the integrable system. As such, we take the sheaf $\mathcal{F}^{\infty/2}$ of semi-infinite forms introduced in Section 2.4.

For any tangent vector X on $\mathcal{P}^D$ let $\rho(X)$ stand for its image under the Kodaira–Spencer mapping. In general, $\rho(X) \in H^1(\Sigma_L, T\Sigma_L)$ where $T\Sigma_L$ is the tangent sheaf on Σ_L. It is shown in Chapter 3 that $\rho(X)$ can be represented by a Krichever–Novikov vector field, and the arising ambiguity is compensated for by passing to the sheaf of coinvariants. Based on those results, below we regard $\rho(X)$ as an element of $\mathcal{V}_L^{\text{reg}} \backslash \mathcal{V}_L / \mathcal{V}_L^{(1)}$ where $\mathcal{V}_L$ is the Lie algebra of Krichever–Novikov vector fields on Σ_L, $\mathcal{V}_L^{(1)}$ is the direct sum of homogeneous subspaces of degree ≥ 0 in $\mathcal{V}_L$, and $\mathcal{V}_L^{\text{reg}} \subset \mathcal{V}_L$ is the subspace of vector fields having zero at P_∞. Both $\mathcal{V}_L^{(1)}$ and $\mathcal{V}_L^{\text{reg}}$ are Lie subalgebras in $\mathcal{V}_L$.

Having represented $\rho(X)$ by a Krichever–Novikov vector field, we may consider $T(\rho(X))$, where T is the Sugawara representation, and, further on, define the high genus Knizhnik–Zamolodchikov connection

$$\nabla_X = \partial_X + T\big(\rho(X)\big).$$

By projective flatness of ∇ we have

$$[\nabla_X, \nabla_Y] = \nabla_{[X,Y]} + \lambda(X,Y) \cdot id,$$

i.e., $X \to \nabla_X$ is a projective representation of the Lie algebra of vector fields on $\mathcal{P}^D$.

Now observe that $\mathcal{P}^D$ is a symplectic manifold with respect to the Krichever–Phong symplectic structure ω [29, 33]. For any $f \in C^\infty(\mathcal{P}^D)$ we can consider the corresponding Hamiltonian vector field X_f. The correspondence $f \to \nabla_{X_f}$ is a representation of the Poisson algebra of classical observables of the Lax integrable system. An easy corollary is as follows. If $\{H_a\}$ is a Poisson commuting family of Hamiltonians labelled by a, then the corresponding Knizhnik–Zamolodchikov operators projectively commute. Moreover, the representation operators of Hamiltonians depending only on the action variables commute.

Having been restricted to the Lie algebra of Hamiltonian vector fields, the representation just introduced is unitary. This claim relies on two facts. First, by the Poincaré theorem the symplectic form and its degrees are absolute integral invariants of Hamiltonian phase flows. Hence the measure on $\mathcal{P}^D$ defined by the volume form $\omega^p/p!$ (where $p = (\dim \mathcal{P}^D)/2$) is invariant with respect to the Hamiltonian flows. For this reason the ∂_X operator is skew-symmetric for any real Hamiltonian X in a certain subspace in $L^2(\mathcal{P}^D, \omega^p/p!)$. Secondly, the Sugawara representation is unitary as soon as the underlying representation of the current algebra possesses this property. For the Virasoro and loop algebras this was proved in [17], and for Krichever–Novikov vector field and commutative current algebras in [27]. The unitarity is considered in detail in Section 6.7.

So far, we have said nothing about the origin of the Sugawara representation. However, to construct it, we must have a representation of a current algebra. Indeed, we have a canonical representation of $\mathcal{L}$ in $\mathcal{F}^{\infty/2}$, but this is not the end of the story. The Sugawara construction (Section 2.7) requires that the current algebra be split to the tensor product of a functional algebra and a finite-dimensional Lie algebra. Krichever–Novikov algebras are of this type, but Lax operator algebras are not. In the next two sections we resolve this problem.

6.2 From Lax operator algebra to commutative Krichever–Novikov algebra

In this section, we canonically associate a commutative Krichever–Novikov algebra to a generic element $L \in \mathcal{L}$. We need it for the Sugawara construction below.

As above, let Ψ be the matrix formed by the canonically normalized left eigenvectors of L (see page 113 for the definition of canonical normalization). Ψ is defined modulo normalization and permutations of its rows (such normalization descends to the left multiplication Ψ by a diagonal matrix). We also consider the diagonal matrix K defined by

$$\Psi L = K\Psi, \tag{6.2.1}$$

i.e., formed by the eigenvalues of L. Let us reproduce here the relations (5.6.5):

$$\begin{aligned} \Psi(z) &= -\frac{\varepsilon\tilde{\beta}\alpha^t\sigma}{z - z_\gamma} + \Psi_0 + \Psi_1(z - z_\gamma) + \cdots, \\ \Psi^{-1}(z) &= \frac{\alpha\tilde{\beta}^t\sigma}{z - z_\gamma} + \tilde{\Psi}_0 + \tilde{\Psi}_1(z - z_\gamma) + \cdots \end{aligned} \tag{6.2.2}$$

in the neighborhood of a γ, and also the relations (5.6.6):

$$\Psi_0\alpha = 0, \quad \varepsilon\alpha^t\sigma\tilde{\Psi}_0 = 0. \tag{6.2.3}$$

Observe that if $\varepsilon = 0$ (i.e., $\mathfrak{g} = \mathfrak{gl}(n)$), then $\nu = \lambda = 0$. Hence $\nu\alpha^t\sigma\tilde{\Psi}_0 = \lambda\alpha^t\sigma\tilde{\Psi}_0 = 0$ as well. For the same reason, $\varepsilon\alpha^t\sigma\alpha = \nu\alpha^t\sigma\alpha = \lambda\alpha^t\sigma\alpha = 0$. As a motivation for the next Lemma 6.1, we recall that by Lemma 5.14 the matrix-valued function K is holomorphic at all γ-points provided (6.2.2) and (6.2.3) hold there. Hence K is a meromorphic diagonal matrix-valued function on Σ holomorphic outside P_i's. If we denote the algebra of scalar-valued functions possessing this property by $\mathcal{A}$, then $K \in \mathfrak{h} \otimes \mathcal{A}$, where $\mathfrak{h} \subset \mathfrak{g}$ is a diagonal (Cartan) subalgebra. $\mathcal{A}$ is a Krichever–Novikov function algebra on Σ (Section 1.1), and $\mathfrak{h} \otimes \mathcal{A}$ is the corresponding Krichever–Novikov current algebra. We obtain only commutative current algebras here. Let $\overline{\mathfrak{h}} = \mathfrak{h} \otimes \mathcal{A}$.

Let $\mathcal{A}_L$ be the Krichever–Novikov function algebra on Σ_L having pre-images of the points P_i (including P_∞) as the collection of poles. Let us take an arbitrary element of $\mathcal{A}_L$ and push it down to Σ as a diagonal matrix h. Every sheet of Σ_L is associated with a certain row of h. At the branching points we may obtain a coincidence of the eigenvalues of h. The order of the diagonal elements of the matrix h will depend on the order of the sheets. This ambiguity is the same as for Ψ. The permutation of rows corresponding to an element w of the Weil group descends to $\Psi \to w\Psi$ (which is easily verified for w to be a transposition), and h transforms as $h \to whw^{-1}$. Thus $L_h = \Psi^{-1}h\Psi$ does not depend on w. The mapping $\mathcal{A}_L \to \overline{\mathfrak{h}}$ is just the direct image of functions from $\mathcal{A}_L$, and L_h is independent of the order of the sheets.

The commutative algebra $\mathcal{A}_{\mathcal{L}}$ is canonically associated with L.

6.3 The representation of $\mathcal{A}_L$

We need a representation of $\mathcal{A}_L$ in order to construct the bundle of conformal blocks and Knizhnik–Zamolodchikov connection in the next section.

Lemma 6.1. *For any $h \in \overline{\mathfrak{h}}$ we have $\Psi^{-1}h\Psi \in \mathcal{L}$.*

Proof. Let us take h in the form

$$h = h_0 + h_1 z + \cdots$$

in a neighborhood of a γ where we set $z_\gamma = 0$ for simplicity. By (6.2.2) we obtain

$$\begin{aligned} L = \Psi^{-1}h\Psi &= \frac{\varepsilon\alpha\tilde{\beta}^t\sigma h_0\tilde{\beta}\alpha^t}{z^2} + \frac{\alpha\tilde{\beta}^t\sigma h_0\Psi_0 + \varepsilon\tilde{\Psi}_0 h_0\tilde{\beta}\alpha^t\sigma}{z} \\ &+ (\alpha\tilde{\beta}^t\sigma h_0\Psi_1 + \alpha\tilde{\beta}^t\sigma h_1\Psi_0 + \varepsilon\alpha\tilde{\beta}^t\sigma h_2\tilde{\beta}^t\alpha^t\sigma + \tilde{\Psi}_0 h_0\Psi_0 \\ &+ \varepsilon\tilde{\Psi}_0 h_1\tilde{\beta}^t\alpha^t\sigma + \varepsilon\tilde{\Psi}_1 h_0\tilde{\beta}^t\alpha^t\sigma) \\ &+ \dots . \end{aligned} \tag{6.3.1}$$

If $\mathfrak{g} = \mathfrak{so}(n)$ then $\sigma^t = \sigma$. We have $(\sigma h_0)^t = h_0^t \sigma^t = -\sigma h_0 \sigma^{-1} \cdot \sigma = -\sigma h_0$, i.e., σh_0 is skew-symmetric. Hence $\tilde{\beta}^t(\sigma h_0)\tilde{\beta} = 0$ and the term with z^{-2} in (6.3.1) vanishes. Observe that for $\mathfrak{g} = \mathfrak{sp}(2n)$ it does not vanish because σh_0 is symmetric. In this case the coefficient is equal to $\varepsilon(\tilde{\beta}^t \sigma h_0 \tilde{\beta})\alpha\alpha^t$ (where $(\tilde{\beta}^t \sigma h_0 \tilde{\beta})$ is a scalar).

Let us consider the term with z^{-1}. We must represent the nominator in the form $(\alpha\widehat{\beta}^t + \varepsilon\widehat{\beta}\alpha^t)\sigma$, i.e., prove that $\alpha\tilde{\beta}^t\sigma h_0 \Psi_0 \sigma^{-1} = (\tilde{\Psi}_0 h_0 \tilde{\beta}\alpha^t)^t$. By $h_0^t = -\sigma h_0 \sigma^{-1}$ it would be sufficient that $\tilde{\Psi}_0^t = -\sigma\Psi_0\sigma^{-1}$. The latter follows from the requirement $(\Psi^{-1})^t = -\sigma\Psi\sigma^{-1}$, i.e., $\Psi \in G$ where G is a classical group corresponding to $\mathfrak{g}$, and which is included in the definition of Ψ above.

Next, let us check the eigenvalue condition on L_0. To this end, let us operate on α by the six-term expression in brackets in (6.3.1). Three summands containing the combination $\varepsilon\alpha^t\sigma$ will vanish by $\varepsilon\alpha^t\sigma\alpha = 0$. Two more summands having Ψ_0 on the right-hand side will vanish by $\Psi_0\alpha = 0$. The remaining summand will give $\alpha(\tilde{\beta}^t\sigma h_0 \Psi_1 \alpha)$. Observe that the expression in brackets is a scalar.

In the symplectic case we must also check the condition $\alpha^t\sigma L_1 \alpha = 0$. This can be done by straightforward computation (already standard) based on the relations (6.2.3) and (5.6.7).

To complete the proof of the theorem we must check that $\Psi^{-1}h\Psi$ is holomorphic except at P_i's and γ_s's. Ψ possesses certain singular points due to the normalization $(\sum \psi_i = 1$ where ψ is a row of $\Psi)$. Let us prove that $\Psi^{-1}h\Psi$ is holomorphic at those points for any $h \in \mathfrak{h}$. By (6.2.1) Ψ is defined up to left multiplication by a diagonal matrix d (corresponding to normalization of the rows of Ψ). But since h and d are both diagonal, $(d\Psi)^{-1}hd\Psi = \Psi^{-1}h\Psi$, and hence $\Psi^{-1}h\Psi$ does not depend on the normalization and can be defined without it. However, then Ψ has no singularities due to normalization. □

The elements $\Psi^{-1}h\Psi \in \mathcal{L}$, $h \in \overline{\mathfrak{h}}$ obviously form a Lie algebra. For a generic L it will be the maximal Lie subalgebra in $\mathcal{L}$ commuting with L (where Ψ is assumed to be the same).

By Lemma 6.1 any representation of $\mathcal{L}$ induces the corresponding representation of $\overline{\mathfrak{h}}$. Since Ψ is meromorphic at P_i's, the mapping $\overline{\mathfrak{h}} \to \mathcal{L}$ just constructed preserves the degree. Hence an almost-graded $\mathfrak{g}$-module induces the almost-graded $\overline{\mathfrak{h}}$-module. Let us recall that Ψ is defined up to left multiplication by a diagonal matrix. For this reason $\overline{\mathfrak{h}}$ needs to be diagonal, otherwise the mapping $\overline{\mathfrak{h}} \to \mathcal{L}$ is not well defined.

Consider the following natural representation of $\mathcal{L}$. Let $\mathcal{F}$ be the space of meromorphic vector-valued functions ψ holomorphic except at $P_1, \dots, P_N, P_\infty$ and γ's, such that

$$\psi(z) = \nu\frac{\alpha}{z} + \psi_0 + \cdots$$

at any point γ. The same space already occurred in Sections 2.2 and 5.6. $\mathcal{F}$ is an almost-graded $\mathcal{L}$-module with respect to the Krichever–Novikov base introduced in

Section 2.3. Consider the semiinfinite degree of this module constructed in Section 2.4. Denote it by $\mathcal{F}^{\infty/2}$. The induced $\overline{\mathfrak{h}}$-module is what we will consider below. This is an *admissible* module in the sense that all of its element annihilate after having been multiply-operated by an element of $\overline{\mathfrak{h}}$ of a positive degree. Moreover, it is generated by a vacuum vector. By the mapping $\mathcal{A}_L \to \overline{\mathfrak{h}}$ constructed above, we also consider $\mathcal{F}^{\infty/2}$ as an $\mathcal{A}_L$-module.

6.4 Sugawara representation

In this chapter we use a "commutative" version of the Sugawara construction, developed in [27], to be applied to $\mathcal{A}_L$. See Section 2.7 for more details.

Remark 6.2. Two problems would arise if we wanted to repeat the Sugawara construction for Lax operator algebras. The first is related to the generalization of the relation $[T(e), u(L)] = u(eL)$ (Theorem 2.21). The right-hand side depends on the choice of a connection $\nabla^{(\omega)} = d + \omega$, and must have the form $[T(e), u(L)] = u(\nabla_e^{(\omega)} L)$ (cf. Section 4.4), but it is not clear how $\nabla^{(\omega)}$ could enter the left-hand side. Another problem is related to the definition of the coefficients $c_m^{jk} = -\operatorname{res}_{P_\infty} \omega^j \omega^k e_m$ involved in the definition of $T(e_m)$, namely, how to make the corresponding 1-form $\omega^j \omega^k e_m$ regular at the γ-points. Another apect of the problem is that, to carry out the Sugawara construction we need the current algebra to be split to a tensor product of a finite-dimensional Lie algebra, and a function algebra, as we already mentioned.

By introducing $\overline{\mathfrak{h}}$ and $\mathcal{A}_L$ we manage with the version of Sugawara construction for Krichever–Novikov algebras, and even for commutative Krichever–Novikov algebras. Unfortunately, the Sugawara construction has never been considered for several outgoing points (which is the case on Σ_L where all preimages of P_∞ are outgoing points), but this can be taken care of using the existing technique (the semiinfinite monomials are considered in [42–44] in this case). Moreover, the Kodaira–Spencer cocycle is a local object (in the topology of $\mathcal{P}^D$), and, as shown in Section 3.1, it is completely defined by its restriction on a neighborhood of a single outgoing point. Hence one can locally fix one of the preimages of P_∞ (i.e., locally fix a sheet of the covering $\Sigma_L \to \Sigma$ over P_∞), and find oneself in the well-described situation.

The role of L is to define an admissible representation of $\overline{\mathfrak{h}}$, $\mathcal{A}_L$. We take $V = \mathcal{F}^{\infty/2}$ as such here. Observe that the Lax operator L is fixed as a function on $\mathcal{L}^D$ and of the spectral parameter, and hence Ψ is uniquely defined up to the equivalence discussed above, and the representation is well defined.

6.5 Conformal blocks and the Knizhnik–Zamolodchikov connection

Let us consider the sheaf of $\mathcal{A}_L$-modules $\mathcal{F}^{\infty/2}$ on $\mathcal{P}^D$. Let $\mathcal{A}_L^{reg} \subset \mathcal{A}_L$ be the subalgebra of functions regular at the preimages (or a locally fixed single preimage; see the discussion in the previous section) of P_∞. The sheaf of quotients $\mathcal{F}^{\infty/2}/\mathcal{A}_L^{reg}\mathcal{F}^{\infty/2}$ on $\mathcal{P}^D$ is called the sheaf of *covariants* (defined by (3.2.4) in this way over a different base).

Let X be a vector field on $\mathcal{P}^D$. By definition,

$$\nabla_X = \partial_X + T\big(\rho(X)\big), \tag{6.5.1}$$

where ρ is the Kodaira–Spencer mapping and T is the Sugawara representation in $\mathcal{F}^{\infty/2}$.

Theorem 6.1. *∇ is a projective flat connection on the sheaf of coinvariants*

$$[\nabla_X, \nabla_Y] = \nabla_{[X,Y]} + \lambda(X,Y)\cdot id$$

where λ is a certain cocycle and id is the identity operator.

In Section 3.4, Theorem 6.1 was formulated and proven in the conformal field theory setup (Proposition 3.14, Theorem 3.18), i.e., for a certain moduli space of Riemann surfaces with marked points and fixed jets of local coordinates at those points. We assert that the situation here is quite similar, and that the proof is the same. In analogy with CFT we refer to the projective flat connection defined by Theorem 6.1 as the *Knizhnik–Zamolodchikov* connection.

The horizontal sections of the sheaf of covariants are called *conformal blocks*.

6.6 The representation of the algebra of Hamiltonian vector fields and commuting Hamiltonians

By Theorem 6.1, $X \to \nabla_X$ is a projective representation of the Lie algebra of vector fields on $\mathcal{P}^D$ in the space of sections of the sheaf of covariants. Denote this representation by ∇. The restriction of ∇ to the subalgebra of Hamiltonian vector fields gives its projective representation.

Theorem 6.2. *If X, Y are Hamiltonian vector fields such that their Hamiltonians Poisson commute, then $[\nabla_X, \nabla_Y] = \lambda(X,Y)\cdot id$. If the Hamiltonians depend only on action variables, then $[\nabla_X, \nabla_Y] = 0$.*

Proof. The projective commutativity immediately follows from Theorem 6.1, since $[X,Y] = 0$.

For a Lax equation, the spectral curve is an integral of motion (more precisely, the coefficients of its equation are integrals). This means that if X is a Hamiltonian vector field such that its Hamiltonian depends only on the action variables, then the complex structure and hence the transition functions are invariant along the phase trajectories of X. Denote by d_L the transition function over L. By its invariance $\partial_X d_L = 0$, hence $\rho(X) = d_L^{-1}\partial_X d_L = 0$, and $\nabla_X = \partial_X$. Let H_X, H_Y be Hamiltonians depending only on the action variables. Then $[\nabla_X, \nabla_Y] = [\partial_X, \partial_Y] = \partial_{[X,Y]}$, and $[X, Y] = 0$. This implies the commutativity of ∇_X, ∇_Y. □

6.7 Unitarity

The goal of the section is to specify a subspace of the representation space and introduce a scalar product there, such that the above representation ∇ becomes unitary.

Recall that in general only real vector field Lie algebras admit unitary representations in the classical sense, i.e., such that the representation operators are skew-Hermitian. For complexifications, a convenient way to define unitarity is as follows [17]. Let $\mathcal{G}$ be a Lie algebra and T its representation in the space V. Consider an antilinear antiinvolution $\dagger$ on $\mathcal{G}$ ($X \to X^\dagger$, $X \in \mathcal{G}$). A Hermitian scalar product in V is called contravariant if $T(X)^\dagger = T(X^\dagger)$, where the $\dagger$ on the left-hand side means the Hermitian conjugation. A pair consisting of T and a contravariant scalar product is called a unitary representation of $\mathcal{G}$. The restriction of T to the Lie subalgebra of the elements such that $X^\dagger = -X$ is unitary in the classical sense.

To construct a contravariant Hermitian scalar product in the space of the representation ∇ we first introduce a point-wise scalar product in the sheaf of covariants, and then integrate it over the phase space $\mathcal{P}^D$ by an invariant volume form.

Over every point in $\mathcal{P}^D$ we introduce a Hermitian scalar product in $\mathcal{F}^{\infty/2}$, declaring semiinfinite monomials with basis entries to be orthonormal [17,27, p. 39]. Below (Theorem 6.3) we prove that it gives a well-defined point-wise scalar product on coinvariants.

By the Poincaré theorem the symplectic form and its degrees are absolute integral invariants of Hamiltonian phase flows. Hence $\omega^p/p!$ defines an invariant measure with respect to Hamiltonian phase flows. Let $\mathcal{L}^2(C, \omega^p/p!)$ be the space of quadratically integrable sections of the sheaf C with respect to that measure, where the square over a point is given by the point-wise scalar product above.

Theorem 6.3. *The representation $\nabla : X \to \nabla_X$ of the Lie algebra of Hamiltonian vector fields on $\mathcal{P}^D$ in the subspace of smooth sections in $\mathcal{L}^2(C, \omega^p/p!)$ is unitary.*

Proof. First let us prove that the point-wise scalar product on $\mathcal{F}^{\infty/2}$ is well defined on coinvariants. The different quasihomogeneous almost-graded components of the module $\mathcal{F}^{\infty/2}$ are orthogonal. The point-wise coinvariants are defined as $\mathcal{V}/\mathfrak{g}^{\text{reg}}\mathcal{V}$. The degrees of monomials occurring in the subspace $\mathfrak{g}^{\text{reg}}\mathcal{V}$ (including summands in linear combinations) are obviously less then those of monomials that form coinvariants. Hence these spaces of monomials are orthogonal, the induced scalar product on the quotient does not depend on the choice of representatives in the equivalence classes, and we get a well-defined point-wise scalar product on coinvariants.

Next let us construct a certain antilinear antiinvolution on the tangent vector fields on $\mathcal{P}^D$. We will push it down from $\mathcal{V}$ by the inverse to the Kodaira–Spencer mapping, making use of the double coset description of the tangent space to the moduli space of curves ($T_L\mathcal{M} = \mathcal{V}_-^{(1)}\backslash\mathcal{V}/\mathcal{V}_+^{(1)}$ where $\mathcal{V}_\pm^{(1)}$ are defined below). Recall from [17, p. 39] that an appropriate antiinvolution in $\mathcal{V}$ is induced by its embedding into the Lie algebra $\mathfrak{a}_\infty$ of infinite matrices with finite number of diagonals. Following [17] we denote this embedding by r. The antilinear antiinvolution in $\mathfrak{a}_\infty$ amounts in $r(e_i) \to -r(e_i)^t$ for the elements of the Krichever–Novikov base, with the consequent antilinear continuation to the complexification. We continue denoting this antiinvolution by $\dagger$. For example, in the case of two marked points $\{P_1, P_\infty\} = \{P_+, P_-\}$ the $\dagger$ sends the subspace $\mathcal{V}_+^{(p)}$ of vector fields regular at P_+ and vanishing there with the order of at least p to the similar subspace $\mathcal{V}_-^{(p)}$ at P_- (for an arbitrary $p \in \mathbb{Z}$). Hence $\mathcal{V}_+^{(p)} \oplus \mathcal{V}_-^{(p)}$ is invariant under the antiinvolution. By Proposition 3.3 and Lemma 3.1 $\mathcal{V}_+^{(1)} \oplus \mathcal{V}_-^{(1)} = \ker\rho^{-1}$ in this case. Hence $\dagger$ is well defined on the tangent space to $\mathcal{P}^D$ at the corresponding point.

For a local vector field X on $\mathcal{P}^D$ by ∂_X we mean the corresponding derivative of local (for example finitary) smooth sections of the sheaf C. For a Hamiltonian X the ∂_X is skew-Hermitian by X-invariance of the measure, hence $\partial_X^\dagger = -\partial_X$. Continuing, for a real Hamiltonian vector field X (i.e., $X^\dagger = -X$) we have $-\partial_X = \partial_{X^\dagger}$, hence $\partial_X^\dagger = \partial_{X^\dagger}$. By complex antilinearity we obtain the same relation for all Hamiltonian vector fields.

Let $\langle\cdot|\cdot\rangle$ be the point-wise scalar product introduced above, and $(\cdot|\cdot) = \frac{1}{p!}\int_{\mathcal{P}^D}\langle\cdot|\cdot\rangle\omega^p$. By [27] $\langle\cdot|\cdot\rangle$ is a contravariant form with respect to the Sugawara representation (in the Abelian case). This implies that the $(\cdot|\cdot)$ possesses the same property, hence $T(\rho(X))^\dagger = T(\rho(X)^\dagger)$. By definition, $\rho(X)^\dagger = \rho(X^\dagger)$. Hence $T(\rho(X))^\dagger = T(\rho(X^\dagger))$.

By the last two paragraphs we have for a Hamiltonian vector field X

$$\partial_X^\dagger = \partial_{X^\dagger} \text{ and } T\big(\rho(X)\big)^\dagger = T\big(\rho(X^\dagger)\big).$$

Hence

$$\Big(\partial_X + T\big(\rho(X)\big)\Big)^\dagger = \Big(\partial_{X^\dagger} + T\big(\rho(X^\dagger)\big)\Big),$$

i.e.,

$$(\nabla_X)^\dagger = \nabla_{X^\dagger}. \qquad \square$$

6.8 Relation to geometric quantization and quantum integrable systems

The subject of this chapter should certainly be discussed in the context of the quantization of Lax integrable systems. This chapter is very similar in spirit to Hitchin [12]. Both are devoted to the problem of a correspondence between an integrable system and a connection on a certain moduli space, in a different setup. In [12] N. J. Hitchin notes two problems of his approach: the taking into account the marked points on Riemann surfaces, and the unitarity of the connection. He points out that the Knizhnik–Zamolodchikov connection could be a solution for the first problem. As shown above, it also resolves the second one.

A large number of works are devoted to quantum integrable systems. Let us give here a brief outline of the construction due to B. Feigin and E. Frenkel [5], A. Beilinson and V. Drinfeld [1]. Denote by $U(\mathfrak{g})_{\varkappa}$ the universal enveloping algebra of $\mathfrak{g}$ on the level $\varkappa$. This is the only case where $U(\mathfrak{g})_{\varkappa}$ has an infinite-dimensional center, namely if $\varkappa$ is equal to the dual Coxeter number for $\mathfrak{g}$. By means of the double-coset construction one can obtain an action of the finite-dimensional quotient space of the center by differential operators in the smooth sections of some vector bundle on the moduli space of holomorphic vector bundles on a Riemann surface. Indeed, $\mathcal{M} = G_{left} \setminus G/G_+$ where $\mathcal{M}$ is the moduli space of holomorphic vector bundles, G is the loop group corresponding to $\mathfrak{g}$, G_+ is its subgroup corresponding to $\overline{\mathfrak{g}}_+$, and G_{left} is a certain subgroup of G_-. The center of $U(\mathfrak{g})_{\varkappa}$ acts on G by casimirs. This action pushes down to $\mathcal{M}$. Not every vector bundle on $\mathcal{M}$ admits a nontrivial space of differential operators. It exists in the more or less unique case of $\sqrt{\mathcal{K}}$ – the bundle of half-forms on $\mathcal{M}$ (where $\mathcal{K}$ is the canonical bundle on $\mathcal{M}$). This construction results in the quantum Hitchin system in the sense that the symbols of the obtained operators are equal to the corresponding classical Hitchin Hamiltonians. Observe that the authors do not quantize the full algebra of observables, but only its commutative subalgebra (see e.g. [1, Sect. 2.2.5]).

Especially detailed information is obtained on quantum Calogero–Moser systems due to the works of A. P. Veselov, A. N. Sergeev, G. Felder, and M. V. Feigin.

In conclusion, let us remind the reader that the results of the present chapter provide a prequantization of the Lax integrable systems with the spectral parameter on a Riemann surface. We prequantize the whole algebra of observables rather than any commutative subalgebra of it.

6.9 Remark on the Seiberg–Witten theory

The Seiberg–Witten theory is a certain ansatz yielding solutions to $\mathcal{N} = 2$ supersymmetric Yang–Mills theories. A complete description of a $\mathcal{N} = 2$ supersymmetric Yang–Mills theory is given by the *partition function* of the form $\widetilde{\mathcal{F}} = \int e^{i\hbar S(\psi)} D\psi$

where S is the Yang–Mills action, ψ runs over gauge fields. A certain low energy limit of $\widetilde{\mathcal{F}}$ is given by the *Seiberg–Witten prepotential* $\mathcal{F}$.

In [56] the following ansatz has been proposed for $\mathcal{F}$. There is a family of pairs (Σ, λ) where Σ is a Riemann surface, λ is a 1-form on it. Let a_i, b_i be the periods of λ with respect to some canonical base of cycles $\{A_i, B_i\}$. Then $\mathcal{F}$ is a function of $\{a_i, b_i\}$ satisfying the following *Seiberg–Witten equations*:

$$b_i = \frac{\partial \mathcal{F}}{\partial a_i}.$$

The following observation [10] relates prepotentials and integrable systems: in all known cases the above family of Riemann surfaces is nothing but the family of the spectral curves of an integrable system, fibered over the phase space of the latter. The corresponding family of 1-forms in our terms is given by[1]

$$\lambda = \varkappa \cdot \varpi,$$

where $\varkappa$ is the spectral function of L, ϖ is the same as in Section 5.5. Then

$$a_i = \oint_{A_i} \varkappa \cdot \varpi, \quad b_i = \oint_{B_i} \varkappa \cdot \varpi$$

are integrals of motion (depend only on the spectrum of Lax operator). They are dependent, so that

$$\sum_i da_i \wedge db_i = 0.$$

In other words, $d(\sum_i b_i da_i) = 0$, hence the 1-form $\sum_i b_i da_i$ has potential $\mathcal{F}$ obviously satisfying the Seiberg–Witten equations.

1 I am obliged to I. M. Krichever for this remark.

Bibliography

[1] Beilinson A., Drinfeld V., Quantization of Hitchin's integrable system and Hecke eigensheaves, preprint.

[2] Bernard D., On the Wess–Zumino–Witten models on the torus, *Nucl. Phys. B,* **303**: 77–93, 1988.

[3] Bernard D., On the Wess-Zumino-Witten models on Riemann surfaces. *Nucl. Phys. B,* **309**: 145–174, 1988.

[4] Bonora L., Rinaldi M., Russo J., Wu K., The Sugawara construction on genus g Riemann surfaces, *Phys. Lett. B*, **208**: 440–446, 1988.

[5] Feigin, B., Frenkel, E., Affine Kac–Moody algebras at the critical level and Gelfand-Dikii algebras. In: A. Tsuchiya, T. Eguchi, M. Jimbo, eds., Infinite analysis, Adv. Ser. in Math. Phys. 16, pp. 197–215. World Scientific, Singapore, 1992.

[6] Feigin B. L., Fuks D. B., Skew-symmetric invariant differential operators on the line and Verma modules over the Virasoro algebra. *Funct. Anal. Appl.,* **16**(2): 114–126, 1982.

[7] Felder G., Wieczerkowski Ch. Conformal blocks on elliptic curves and Knizhnik–Zamolodchikov–Bernard equations. *Comm. Math. Phys.*, **176**: 133–161, 1996.

[8] Friedan D., Shenker S., The analitic geometry of the two-dimensional conformal field theory. *Nuclear Phys. B*, **281**: 509–545, 1987.

[9] Garland H., The arithmetic theory of loop groups, *Publ. Math. IHES*, **52**: 5–136, 1980.

[10] Gorsky A., Krichever I., Marshakov A., Mironov A., Morozov A., Integrability and Seiberg–Witten exact solution, `Arx.math.9505035`.

[11] Grinevich P. G., Orlov A.Yu., Variations of the complex structure of Riemann surfaces by vector fields on the circle, and objects of the KP theory. The Krichever–Novikov problem of action on the Baker-Akhieser functions. *Funct. Anal. Appl.*, **24**(1): 61–63, 1991.

[12] Hitchin N., Flat connections and geometric quantization, *Commun. Math. Phys.*, **131**: 347–380, 1990.

[13] D'Hocker E., Phong D. H. Calogero-Moser Lax pairs with spectral parameter for general Lie algebras, `Hep-th/9804124`.

[14] Ivanov D., Knizhnik–Zamolodchikov–Bernard equations as a quantization of nostationary Hitchin systems, `Hep-th/9610207`.

[15] Kac V. G., Infinite-dimensional Lie algebras, 3rd edn., Cambridge University Press, 440 pp, 1990.

[16] Kac V. G., Peterson D. H., Spin and wedge representations of infinite-dimensional Lie algebras and groups. *Proc. Natl. Acad. Sci. USA*, pp. 3308–3312, 1981

[17] Kac V. G., Raina A. K., Highest Weight Representations of Infinite Dimensional Lie Algebras, Adv. Ser. in Math. Physics 2, World Scientific, 1987.

[18] Kassel Ch., Kähler differentials and coverings of complex simple Lie algebras extended over a commutative algebra, *J. Pure Appl. Algebra*, **34**: 265–275, 1984.

[19] Knizhnik V. G., Zamolodchikov A. B., Current algebra and Wess-Zumino model in two dimensions, *Nucl. Phys. B*, **247**: 83–103, 1984.

[20] Kodaira K., Complex manifolds and deformation of complex structures, Springer-Verlag, 1985.

[21] Kontsevich M., Virasoro algebra and Teichmüller space, *Funct. Anal. Appl.*, **21**(2): 156–157, 1987.

[22] Krichever I.M., Algebraic curves and nonlinear difference equations. *Russ. Math. Surv.*, **33**(4): 255–256

[23] Krichever I, 1978. M., Novikov S.P., Holomorphic bundles over Riemann surfaces and the Kadomtsev–Petviashvili equation. I. *Funct. Anal. and Appl.*, **12**(4): 276–286, 1978.

[24] Krichever I. M., Elliptic solutions to the Kadomtsev-Petviashvili equation and integrable systems of particles, *Fuct. Analysis and Appl.*, **14**(4): 45–54, 1980.

[25] Krichever I. M., Novikov S.P., Holomorphic bundles over algebraic curves and non-linear equations. *Russ. Math. Surv.*, **35**(6): 53–79, 1980.

[26] Krichever I. M., Novikov S.P., Algebras of Virasoro type, Riemann surfaces and structures of the theory of solitons, *Funct. Anal. and Appl.*, **21**(2): 126–142, 1987.

[27] Krichever I. M., Novikov S.P., Algebras of Virasoro type, Riemann surfaces and strings in Minkowsky space, *Funct. Anal. and Appl.*, **21**(4): 294–307, 1987.

[28] Krichever I. M., Novikov S.P., Algebras of Virasoro type, energy-momentum tensor, and decomposition operators on Riemann surfaces, *Funct. Anal. and Appl.*, 1, **23** (1989), 19–33.

[29] Krichever I. M., Phong D.H., Symplectic forms in the theory of solitons. Surveys in Differential Geometry IV, pp. 239–313, 1998. International press, C. L. Terng, K. Ulenbeck, eds. `Hep-th9708170`.

[30] Krichever I. M., Novikov S.P., Holomorphic bundles and scalar difference operators: one-point constructions. *Russ. Math. Surv.*, **55**(1): 180–181, 2000.

[31] Krichever I. M., Novikov S.P., Holomorphic bundles and commuting difference operators. Two-point constructions, *Russ. Math. Surv.*, **55**(3): 586–588, 2000.

[32] Krichever I. M., Isomonodromy equation on algebraic curves, canonical transformations and Witham equations, `Hep-th/0112096`.

[33] Krichever, I. M., Vector bundles and Lax equations on algebraic curves, *Comm. Math. Phys.*, **229**: 229–269, 2002. Hep-th/0108110.

[34] Krichever I. M., Novikov S.P., Two-dimensionalized Toda lattice, commuting difference operators, and holomorphic bundles. *Russ. Math. Surv.*, **58**(3): 473–510, 2003.

[35] Krichever I. M., Sheinman O.K., Lax operator algebras. *Funct. Anal. i Prilozhen.*, **41**(4): 46–59, 2007. math.RT/0701648.

[36] Levin A., Olshanetsky M., Non-autonomous Hamiltonian systems related to highest Hitchin integrals. Proceedings of the "International Seminar on Integrable systems". In memoriam Mikail V. Saveliev. Bonn, February, 1999. ArXiv:math-ph/9904023.

[37] Olshanetsky M. A.. Generalized Hitchin Systems and the Knizhnik–Zamolodchikov–Bernard Equation on Ellipic Curves, *Lett, Math, Phys.*, **42**(1):, 59–71, 1997.

[38] Polyakov A., Quantum geometry of bosonic strings, *Phys. Lett. B* **103**: 207, 1981.

[39] Ruffing A., Deck Th., Schlichenmaier M., String branchings on complex tori and algebraic representations of generalized Krichever–Novikov algebras. *Lett. Math. Phys.*, **26**: 23–32, 1992.

[40] Sadov V. A., Bases on multipunctured Riemann surfaces and interacting strings amplitudes, *Commun. Math. Phys.*, **136**: 585–597, 1991.

[41] Schlichenmaier M., ,An introduction to Riemann surfaces, algebraic curves and moduli spaces, Lecture Notes in Physics 322, Springer, Berlin, Heidelberg, New York, 1989.

[42] Schlichenmaier M., Krichever–Novikov algebras for more than two points, *Lett. Math.l Phys.*, **19**: 151–165, 1990.

[43] Schlichenmaier M., Krichever–Novikov algebras for more than two points: explicit generators, *Lett. Math. Phys.*, **19**: 327–336, 1990.

[44] Schlichenmaier M., Central extensions and semi-infinite wedge representations of Krichever–Novikov algebras for more than two points, *Lett. Math. Phys.*, **20**: 33–46, 1990.

[45] Schlichenmaier M., Verallgemeinerte Krichever–Novikov Algebren und deren Darstellungen. Ph.D. thesis, Universität Mannheim, 1990.

[46] Schlichenmaier M., Degenerations of generalized Krichever–Novikov algebras on tori, *J. Math. Phys.*, **34**: 3809–3824, 1993.

[47] Schlichenmaier M., Differential operator algebras on compact Riemann surfaces. Generalized Symmetries in Physics (Clausthal 1993, Germany) (H.-D. Doebner, V.K. Dobrev, A.G. Ushveridze, eds.), World Scientific, 1994, pp. 425–434.

[48] Schlichenmaier M., Local cocycles and central extensions for multi-point algebras of Krichever–Novikov type, *J. Reine Angew, Math.*, **559**: 53–94, 2003.

[49] Schlichenmaier M., Higher genus affine algebras of Krichever–Novikov type, *Moscow Math. J.*, **4**(3): 1395–1427, 2003. math.QA/0210360.

[50] Schlichenmaier M., Higher genus infinite–dimensional conformal Lie algebras and their applications, De Gruyter, forthcoming.

[51] Schlichenmaier M., Sheinman O. K., Sugawara construction and Casimir operators for Krichever–Novikov algebras, *J. Math. Sci.*, **92**: 3807–3834, 1998. math.QA/9512016.

[52] Schlichenmaier M., Sheinman O. K. ,Wess–Zumino–Witten–Novikov theory, Knizhnik–Zamolodchikov equations, and Krichever–Novikov algebras, *Russ. Math. Surv.*, **54**(1): 213–249, 1999.

[53] Schlichenmaier M., Sheinman O. K., Knizhnik–Zamolodchikov equations for positive genus and Krichever–Novikov algebras, *Russ. Math. Surv.*, **59**(4): 737–770, 2004.

[54] Schlichenmaier M., Sheinman O. K., Central extensions of Lax operator algebras, *Russ. Math. Surv.*, **63**(4): 727–766, 2008. `ArXiv:0711.4688`.

[55] Schottenloher M., A Mathematical introduction to Conformal Field Theory, Second edition, 249 pp, Lecture Notes in Physics 759, Springer, 2008.

[56] Seiberg N., Witten E., Ellectric-magnetic duality, monopole condensation, and coinfiment in $\mathcal{N} = 2$ supersymmetric Yang–Mills theory. `Arxiv:hep-th/9407087`.

[57] Sheinman O. K., Elliptic affine Lie algebras, *Funct. Anal. and Appl.*, **24**(3): 210–219, 1990.

[58] Sheinman O. K., Highest weight modules over certain quasigraded Lie algebras on elliptic curves, *Funct. Anal. and Appl.*, **26**(3): 203–208, 1992.

[59] Sheinman O. K., Affine Lie algebras on Riemann surfaces,. *Funct. Anal. and Appl.*, **27**(4): 266–272, 1993.

[60] Sheinman O. K., Modules with highest weight for affine Lie algebras on Riemann surfaces, *Funct. Anal. and Appl.*, **29**(1): 44–55, 1995.

[61] Sheinman O. K., Representations of Krichever–Novikov algebras, Topics in topology and mathematical physics, Novikov, S.P. (ed.), pp. 185–197, American Mathematical Society, Providence, RI, USA, 1995.

[62] Sheinman O. K., The Fermion Model of Representations of Affine Krichever–Novikov Algebras, *Funct. Anal. and Appl.*, **35**(3): 209–219, 2001.

[63] Sheinman O. K., Second order casimirs for the affine Krichever–Novikov algebras $\widehat{\mathfrak{gl}}_{g,2}$ and $\widehat{\mathfrak{sl}}_{g,2}$, Moscow Math. J., **1**(4): 605–628, 2001. `math.RT/0109001`.

[64] Sheinman O. K., Second-order Casimir operators for the affine Krichever–Novikov algebras $\widehat{\mathfrak{gl}_{g,2}}$ and $\widehat{\mathfrak{sl}_{g,2}}$, *Russ. Math. Surv.*, **56**(5): 986–987, 2001.

[65] Sheinman O. K., Projective flat connections on the moduli space of Riemann surfaces and Knizhnik–Zamolodchikov equations, *Proc. Steklov Inst. Math., Non-linear dynamics*, **251**: 293–304, 2005.

[66] Sheinman O. K., Krichever–Novikov algebras and their representations, *Contemporary Mathematics*, **391**: 313–321, 2005.

[67] Sheinman O. K., Krichever–Novikov algebras, their representations and applications. Geometry, Topology and Mathematical Physics. S. P. Novikov's seminar 2002–2003. V. M. Buchstaber, I. M. Krichever, eds. AMS translations, Ser. 2, vol. 212. 2004, 297–316. `mathRT/0304020`.

[68] Sheinman O. K., Krichever–Novikov algebras, their representations and applications in geometry and mathematical physics. In: Contemporary mathematical problems, Steklov Mathematical Institute publications 10, 2007, 142 p. (in Russian).

[69] Sheinman O. K., On certain current algebras related to finite-zone integration. In: Geometry, Topology and Mathematical Physics. S. P. Novikov's Seminar 2004–2008, V. M. Buchstaber, I. M. Krichever, eds., AMS Translations, Ser. 2, vol. 224, 2008.

[70] Sheinman O. K., Lax operator algebras and integrable hierarchies. In: Proc. of the Steklov Mathematical Institute, vol. 263, 2008.

[71] Sheinman O. K., Lax operator algebras and Hamiltonian integrable hierarchies, *Russ. Math. Surv.*, **66**(1): 145–171, 2011. `Arxiv.math.0910.4173`.

[72] Sheinman O. K., Lax equations and Knizhnik–Zamolodchikov connection. Proc. Workshop on Geometric Methods in Physics. Birkhauser, 2012, 9 p. `ArXiv1009.4706`.

[73] Sugawara H., A field theory of currents, *Phys. Rev.*, **176**: 2019–2025, 1968.

[74] Tsuchiya A., Ueno K., Yamada Y., Conformal field theory on universal family of stable curves with gauge symmetries, *Adv. Stud. Pure Math.*, **19**: 459–566, 1989.

[75] Tyurin, A. N., Classification of vector bundles over an arbitrary genus algebraic curve, *Soviet Izvestia, ser. Math.*, **29**: 657–688, 1965.

[76] Tyurin, A., Quantization, classical and quantum field theory and theta functions, CRM Monograph Series 21, American Mathematical Society, Providence, RI, 2003, 136 p.

[77] Ueno K., Introduction to conformal field theory with gauge symmetries. Geometry and Physics, In: Proceed. Aarhus conference 1995, Andersen J. E. et al. (eds.), pp. 603–745, Marcel Dekker, 1997.

Notation

$C_{\widetilde{W}}, C_{\widetilde{W}^{(k,p)}}$	sheaves of conformal blocs, 63
D	effective divisor supported at $\{P_1, \dots, \mathcal{P}_N\}$, 109
$H^1\big(\Sigma, T_\Sigma(-(k+1)S$ $-(p+1)P_\infty)\big)$	cohomology with coefficients in the sheaf, 57
$T_L \mathcal{L}^D$	tangent space at $\mathcal{L} \in \mathcal{L}^D$, 110
$V_{\widetilde{W}}, V_{\widetilde{W}^{(k,p)}}$	sheaves of Verma modules, 63
$\mathcal{A}_{\widetilde{W}}, \mathcal{V}_{\widetilde{W}}, \widehat{\mathcal{V}}_{\widetilde{W}}, \overline{\mathfrak{g}}_{\widetilde{W}}, \widehat{\mathfrak{g}}_{\widetilde{W}}$	sheaves of the Lie algebras, 62
$\mathcal{L}$	Lax operator algebra, 91
$\mathcal{L}^D \in \mathcal{L}$	invariant subspace of integrable hierarchy, 109
$\mathrm{T}_{\tilde{b}^{(k,p)}}$	tangent space at the point $\tilde{b}^{(k,p)}$, 57
$\mathcal{V}$	Krichever–Novikov algebra of vector fields, 2
$\widetilde{W}$	open subset of the moduli space $\mathcal{M}^{(1,0)}_{g,N+1}$, 62
$\widetilde{W}^{(k,p)}$	open subset of the moduli space $\mathcal{M}^{(k,p)}_{g,N+1}$, 63
$\overline{\mathfrak{g}}$	algebra of currents
	– of the Krichever–Novikov type, 2
	– of the Lax operator type, 91
$\widehat{\mathfrak{g}}$	affine Krichever–Novikov algebra, 13
$\tilde{b}^{(k,p)}$	point of the moduli space $\mathcal{M}^{(k,p)}_{g,N+1}$, 56
$\mathcal{M}^{(k,p)}_{g,N+1}$	the moduli space, 56
$\mathcal{F}$	space of functions related to Tyurin parameters, natural $\mathcal{L}$-module, 119, 134
$\mathcal{F}$	space of functions related to Tyurin parameters, natural $\mathcal{L}$-module, 22
$\mathfrak{g}^\diamond$	auxiliary Lie algebra, range of values of M-operators, 109
∇_X	Knizhnik–Zamolodchikov covariant derivative along the vector field X, 67
$\widetilde{D}$	auxiliary divisor, 112

Index